AN INTRODUCTION TO
FLUID MECHANICS AND
HEAT TRANSFER

AN INTRODUCTION TO
FLUID MECHANICS AND HEAT TRANSFER

WITH APPLICATIONS IN CHEMICAL & MECHANICAL PROCESS ENGINEERING

BY

J. M. KAY

M.A., Ph.D., A.M.I.Mech.E., A.M.I.Chem.E.

*Professor of Nuclear Power in the University of London
at the Imperial College of Science and Technology*

CAMBRIDGE
AT THE UNIVERSITY PRESS
1957

PUBLISHED BY
THE SYNDICS OF THE CAMBRIDGE UNIVERSITY PRESS
Bentley House, 200 Euston Road, London, N.W.1
American Branch: 32 East 57th Street, New York 22, N.Y.

Made in Great Britain at the Pitman Press, Bath

PREFACE

This book is based on lectures given in the Department of Chemical Engineering at Cambridge from 1948 to 1952. The main emphasis is on the fundamental principles of fluid mechanics and heat transfer, but these principles are related wherever possible to actual engineering problems. No attempt has been made, however, to compile a designer's handbook. It is assumed that the reader will have access to other specialized textbooks giving the numerical data and formulae appropriate to various fields of application.

The object has been to cover the more important aspects of fluid flow and the transfer of heat and mass from the point of view of chemical and mechanical process engineering. The book falls into three distinct sections. Chapters 1–11 constitute an elementary introduction to the subject. Chapters 12–17 form a hard core of basic theory. Chapters 18–22 are concerned with certain special applications which are of interest in process engineering. A number of worked examples are included in the text and some additional numerical examples have been added at the end of the book. Many of these are taken from examination papers set for the Chemical Engineering Tripos.

The author wishes to acknowledge the great assistance given to him by Professor E. S. Sellers in reading through the manuscript and offering advice on many points. He also wishes to acknowledge his debt to Professor T. R. C. Fox for guidance on many aspects of the subject, and to Mr R. Vaux for help in correcting the proofs.

J. M. K.

CAMBRIDGE

CONTENTS

Contents

ix

x   *Contents*

Contents   xiii

Appendices

LIST OF SYMBOLS

a	Cross-sectional area of stream tube, radius of a pipe, velocity of sound
A	Area
b	Linear dimension, e.g. breadth of pump impeller
c	A coefficient, constant, concentration
c_A	Concentration of component (A)
c_f	Skin friction coefficient
c_p	Specific heat at constant pressure
c_v	Specific heat at constant volume, velocity coefficient
C	A constant, discharge coefficient
C_A	Molal concentration of component (A)
C_D	Drag coefficient
d	Diameter
D	Diameter
$\mathscr{D}$	Diffusion coefficient
e	Exponential constant
E	Internal energy
f	Flow component of velocity in a pump
F	Force
g	Gravitational acceleration
G	Mass velocity $= \rho u$
G_A	Mass transfer rate for (A)
h	Height, head, heat transfer factor
H	Enthalpy, dimension of heat, head
$\mathbf{i}$	Unit vector in x direction
$\mathbf{j}$	Unit vector in y direction
J	Mechanical equivalent of heat
k	A coefficient, thermal conductivity
$\mathbf{k}$	Unit vector in z direction
K	A constant
l	Length
l'	Mixing length
L	Dimension of length, latent heat
m	Mass, hydraulic mean depth
M	Mass, dimension of mass, molecular weight, momentum

n	Number
$\mathbf{n}$	Unit vector normal
N	Number, rotational speed
N_A	Molal transfer rate for (A)
O	Order of magnitude
p	Pressure
P	Power
q	Heat flux
Q	Volumetric flow, heat quantity, rate of heat supply
r	Radius
$\mathbf{r}$	Position vector
R	Radius, gas constant
$\bar{R}$	Universal gas constant per mole
s	Distance
S	Surface area
t	Time
T	Temperature, torque, dimension of time
u, v, w	Velocity components in x, y, z directions
$\mathbf{u}$	Vector velocity
u_m	Mean velocity in a pipe
U	Velocity
v	Velocity, specific volume
$\tilde{v}$	r.m.s. velocity in turbulent flow
V	Volume
w	Velocity of whirl
W	Work
x, y, z	Space coordinates
α	An angle, coefficient, thermal diffusivity
β	An angle, coefficient
γ	Ratio of specific heats c_p/c_v
δ	Element of length, thickness of boundary layer
δ^*	Displacement thickness of boundary layer
Δ	Incremental quantity
ε	Roughness size, fraction void
η	Efficiency
θ	Angle, temperature, momentum thickness of boundary layer
Θ	Dimension of temperature
λ	A parameter, depth of flow in open channel

μ	Coefficient of viscosity
ν	Kinematic viscosity, molecular density
Π	Dimensionless ratio
ρ	Density
σ	Surface tension
τ	Skin friction, shearing stress
τ_0	Skin friction at surface or boundary
ϕ	Angle, coefficient, thermal thickness of boundary layer
ω	Angular velocity
Ω	Angular velocity
∇	Vector operator 'nabla'

DIMENSIONLESS RATIOS

Fr Froude number $\dfrac{U^2}{gL}$

Gr Grashof number $\dfrac{\beta g \Delta \theta L^3}{v^2}$

M Mach number U/a

Nu Nusselt number $\dfrac{hd}{k}$

Pe Peclet number $\dfrac{\rho U c_p d}{\kappa}$ or $\dfrac{Ud}{\alpha}$

Pr Prandtl number $\dfrac{\mu c_p}{\kappa}$ or $\dfrac{v}{\alpha}$

Re Reynolds number $\dfrac{\rho Ud}{\mu}$ or $\dfrac{Ud}{v}$

Sc Schmidt number $\dfrac{v}{\mathscr{D}}$

St Stanton number $\dfrac{h}{\rho U c_p}$

We Weber number $\dfrac{\rho U^2 L}{\sigma}$

CHAPTER 1

INTRODUCTION AND DEFINITIONS

Fluid mechanics is a subject of widespread interest to engineers. The study of *hydraulics* as a science is associated historically with the growth of civil engineering and of naval architecture during the nineteenth century. The development of mechanical and chemical engineering during the past 50 years, however, has given an added stimulus to the wider study of *fluid mechanics*, so that it now ranks as one of the most important basic subjects in engineering science. This book is written primarily from the point of view of chemical and mechanical process engineering. In chemical engineering we are mainly concerned with the flow of gases and liquids through different types of plant and equipment and with the associated energy and material transfer phenomena. It is therefore natural to link up fluid mechanics with the study of *heat and mass transfer*. This approach may also be useful to many mechanical engineers.

1.1. Liquids and gases

Fluids include both liquids and gases. The feature that distinguishes a fluid from a solid is the inability of a fluid, while remaining in a state of equilibrium, to resist shearing forces.

The distinction between a liquid and a gas is readily appreciated but is difficult to define in a concise manner. A definite mass of a particular liquid will have a definite volume which will vary only slightly with temperature and pressure. The volume of a definite mass of a gas, on the other hand, is almost infinitely variable, and its dependence on pressure and temperature will be given by the equation of state. The so-called permanent gases such as nitrogen and oxygen follow more or less closely the perfect gas law $pV = mRT$. A liquid will exhibit a free surface if a larger space is available than that required to contain its volume. A gas, on the other hand, will fill any space available to it, the pressure adjusting itself accordingly.

The physical properties of gases can be explained on a molecular picture by the kinetic theory of gases[1],† but there is as yet no adequate kinetic theory of liquids. It might be thought that it would be necessary, for engineering purposes, to develop the mechanics of liquids and gases as separate subjects, but this is not

† Small figures in parentheses refer to References at end of chapter.

in fact the case. Provided we are dealing with problems of flow in which density changes are small we can regard gases as being incompressible, and the theory of *incompressible flow* is identical for both liquids and gases. The assumption of constant density for a gas or vapour will break down, however, if we have large thermal effects involving appreciable changes of temperature or if we have very high velocities. These conditions apply, for instance, in the flow of steam through a boiler plant and turbine, in certain types of heat exchanger, and in the high-speed flight of aeroplanes or projectiles. The theory of *compressible flow* has been developed to cover these cases.

1.2. Hydrostatic pressure

It is assumed that the reader will be familiar with the principles of hydrostatics. It will therefore only be necessary to recall that for a fluid at rest the pressure at a point is the same in all directions. This follows directly from the absence of shear stresses. In other words, for a fluid in equilibrium, the internal state of stress at any point is specified completely by the pressure. It will also be recalled that, for equilibrium under gravity, the pressure is uniform over any horizontal cross-section, but that it varies with height or depth in accordance with the relation

$$\mathrm{d}p = -\rho g \mathrm{d}h, \qquad (1.2.1)$$

where ρ is the density or mass per unit volume, g is the gravitational acceleration, and h is the height or depth measured positive upwards. Some writers use the symbol w for the specific weight in place of the product ρg. Normally when we refer to pressure we shall mean the absolute pressure. Sometimes, however, it is convenient to quote gauge pressure, i.e. the value measured relative to atmospheric pressure.

Note that the *density* ρ is mass per unit volume and has dimensions ML^{-3}. It is normally measured in pounds per cu.ft. With g expressed in ft./sec.^{2} and h in ft., equation (1.2.1) will give the pressure in absolute units, i.e. in poundals/sq.ft.

The *specific weight* w, on the other hand, is weight per unit volume and has dimensions $ML^{-2}T^{-2}$ or FL^{-3} (see footnote† on dimensions). If w is quoted in absolute units (poundals/cu.ft.) it will be numerically equal to the product ρg, and the pressure in equation (1.2.1) will still be in absolute units (poundals/sq.ft.). It is more usual, however,

† *Footnote on dimensions.* Mechanical quantities are normally expressed in terms of the three fundamental dimensions of mass M, length L and time T. On this system force F is a derived quantity having dimensions given by the

to quote w in lb.wt./cu.ft. It will then be numerically, but not dimensionally, equal to the density ρ. Using the relation

$$\mathrm{d}p = -\,w\mathrm{d}h,$$

the pressure will now be given directly in lb.wt./sq.ft.

1.3. Fluids in motion

Although a fluid which is at rest cannot sustain shearing forces, this is no longer the case with a fluid in motion. The existence of shearing stresses in a moving fluid is associated with the physical property of *viscosity*. The action of viscosity will be discussed in the next section, but, first of all, it will be necessary to draw attention to another experimental fact.

A fluid does not slip at a solid boundary. When a fluid is flowing parallel to a solid surface the particles of fluid immediately adjacent to the boundary are in fact at rest. There must therefore be a region in the flow in which the velocity rises from zero at the boundary to the full value of the undisturbed velocity of the main stream. In some cases this region is quite thin and is known as a *boundary layer*. In other cases, for instance, with flow in a pipe at some distance from the entry, the boundary region may extend for an appreciable radial distance away from the wall towards the centre of the pipe. Wherever we have large transverse velocity gradients in this way the effect of viscosity is important and can never safely be neglected.

---

product of mass and acceleration, i.e. MLT^{-2}. Three sets of units are in common use as shown in the following table:

	M	L	T	$F = MLT^{-2}$ absolute unit	practical unit
English units	pound	foot	second	poundal	lb. weight
Continental	gram	centimetre	second	dyne	g. weight
MKS	kilogram	metre	second	newton	kg. weight

The magnitude of the practical unit of force in each case is equal to that of the absolute unit multiplied by the *numerical* value of the gravitational acceleration.

On the American system, however, force, length and time are taken as the three fundamental dimensions. Mass M must then be regarded as a derived quantity having dimensions $FL^{-1}T^2$. The following units are employed:

	F	L	T	M
American units	pound	foot	second	slug

The American unit of force, the pound, is numerically identical with the English practical unit of force, the pound weight. Confusion sometimes arises because of the use of the word 'pound' in two different senses, as a unit of mass in the English system, and as a unit of force in the American system. The American unit of mass, the slug, is defined as the mass which experiences unit acceleration (1 ft./sec.²) when acted on by 1 pound force. The slug is therefore 32·2 times the size of the English pound mass.

In certain cases, however, particularly with flow past aerofoils and other thin streamlined shapes, transverse velocity gradients are relatively small outside the boundary region, and the effect of viscosity on the main flow is almost negligible. A very extensive theory has been built up to cover the flow of *ideal* or *inviscid fluids* in which the action of viscosity is ignored altogether. This theory is of some importance to aeronautical engineers and, because it lends itself to rigorous mathematical treatment, it has been pursued to great lengths. Ideal fluid theory, however, is of very limited application outside the realm of aerofoil design. We shall be much more concerned with the behaviour of *real fluids*, particularly in pipes and channels and close to solid boundaries, and with the associated problems of flow resistance and transfer phenomena.

We shall need a mathematical notation in order to describe fluid motion. It will be convenient to take position coordinates, x, y, z, to locate a fixed point in the flow. We can then say that for *steady flow* (i.e. in which the flow pattern does not vary with time) the velocity components u, v, w, measured in the x, y and z directions respectively, are functions of the coordinates x, y, z, i.e.

$$u = f(x, y, z), \quad \text{etc.}$$

Vector notation enables us to put the thing more concisely. If **u** is the velocity vector (components u, v, w), and if **r** is the position vector (components x, y, z), we can say that for *steady flow*

$$\mathbf{u} = f(\mathbf{r}), \tag{1.3.1}$$

and that for *unsteady flow* (i.e. flow varying with time t)

$$\mathbf{u} = f(\mathbf{r}, t). \tag{1.3.2}$$

The notation is illustrated diagrammatically in fig. 1 for a two-dimensional case. The velocity vector **u** is shown for the point P whose position vector is **r**.

It is useful in steady flow to picture a set of lines drawn so as to coincide with the velocity vectors at every point. These are known as *streamlines*. By definition there can be no flow across a streamline. If the flow is unsteady, the streamline picture drawn for a particular instant of time would give us in effect an instantaneous photograph of the flow.

We have already noted that in certain regions of flow in the neighbourhood of solid boundaries, particularly flow in pipes and in boundary layers, velocity gradients are large and viscosity plays an important part. It is observed experimentally that in such regions we may have two entirely different types of motion. Under certain conditions the flow will be regular or *laminar*, and if, for instance, we introduce a thin filament of coloured fluid it will remain as a thin filament and will in fact mark out visually a streamline in the flow. We call motion of this kind *laminar flow*. Under other conditions, however, rapid diffusion or dispersion of the coloured material will occur and the flow appears to be of an irregular nature with transverse eddies. This is known as *turbulent flow*, and it may be regarded as an unsteady fluctuating velocity distribution superimposed on a steady mean-flow pattern. In most practical cases of flow through pipe-lines, heat exchangers, pumps, compressors and turbines, the flow is predominantly turbulent. The study of turbulent

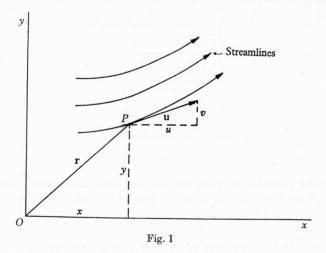

Fig. 1

flow will therefore be of particular importance to us. It must be admitted, however, that there is as yet no completely satisfactory theory of turbulent flow that can be applied to problems of engineering interest.

1.4. Viscosity

Shear stresses in moving fluids are related to velocity gradients or rate-of-strain components in much the same way that stresses in elastic solids are related to strain components. A full discussion of these shear stresses will be given in a later chapter, but let us first take a simple case of flow parallel to a plane wall as in fig. 2 in which we need only consider the transverse velocity gradient. We can say that the velocity u is a function only of y, the distance measured normally to the solid wall, and that the value of u ranges from zero at $y = 0$ to u_1, the undisturbed velocity of the main stream at $y = y_1$. The shear stress at any point in the flow is given by

$$\tau = \mu \frac{du}{dy}, \qquad (1.4.1)$$

where μ is the coefficient of viscosity for the fluid. It will be seen that for the case shown in fig. 2 the shear stress increases as the

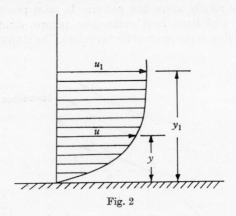

Fig. 2

velocity gradient steepens towards the wall. The highest value of the shear stress is actually at the wall and is represented by

$$\tau_0 = \mu (du/dy)_0.$$

Note that the *coefficient of viscosity* μ has dimensions

$$\frac{M}{LT} \quad \text{or} \quad \frac{FT}{L^2},$$

and it may be measured in accordance with any of the following three systems:

British	$\dfrac{\text{lb. (mass)}}{\text{ft.sec.}}$ or	$\dfrac{\text{poundal sec.}}{\text{sq.ft.}}$
American	$\dfrac{\text{slugs}}{\text{ft.sec.}}$ or	$\dfrac{\text{pound (force) sec.}}{\text{sq.ft.}}$
Continental	$\dfrac{\text{g.}}{\text{cm.sec.}}$ or	$\dfrac{\text{dyne sec.}}{\text{sq.cm.}}$

The continental unit is the poise, but a more useful size of unit is the centipoise equal to one-hundredth part of a poise. The numerical conversion from one system to another is as follows:

1 poise $= 0.0672$ British unit $= 0.00209$ American unit
g./cm.sec.   lb./ft.sec.   slug/ft.sec.

or pound (force) sec./sq.ft.

It is frequently more convenient to make use of the ratio of the coefficient of viscosity to the density. This is known as the *kinematic viscosity* and is denoted by the symbol v. Since v is equal to μ/ρ it must have dimensions L^2/T, and can therefore be measured either in sq.ft./sec. or sq.cm./sec.

The coefficient of viscosity of a gas increases with temperature, and this can be explained in terms of the kinetic theory of gases (see, for instance, Jeans, *Kinetic Theory of Gases*, Chapter VI). The coefficient of viscosity of a liquid, on the other hand, decreases with rise of temperature.

Most fluids behave in the manner described, with shear stresses being proportional to velocity gradients, and are known as 'Newtonian fluids'. There are some fluids, however, in particular, certain slurries and colloidal suspensions, which do not exhibit this simple proportionality between shear stress and velocity gradient. These are known as 'non-Newtonian liquids'.

1.5. Temperature and heat

It is one of the difficulties of the science of thermodynamics that it is impossible to give a satisfactory definition of temperature without bringing in the concept of *heat*, and that it is equally impossible to give a satisfactory definition of heat without bringing in the concept of *temperature*. This is not a textbook of thermodynamics, and it will not therefore be appropriate to launch ourselves into the subtleties of thermodynamic argument. It will be sufficient to note that it is usual to start with the idea of *thermal equilibrium*. Two bodies are said to be at the same temperature if they remain in thermal equilibrium when brought into contact with each other. Scales of temperature are then defined in terms of certain measurable physical properties such as the volumetric expansion of mercury. It is shown in textbooks of thermodynamics that one particular scale of temperature, the perfect-gas scale, has special significance because it is identical with an absolute thermodynamic scale of temperature based on the second law of thermodynamics.

Heat is said to pass from one system to another if the two systems are initially at different temperatures and are then placed in contact with each other. The quantity of heat passing to a system is measured in terms of the product of mass, specific heat and temperature rise. Note that heat is a transient thing and that we can really only talk about *heat transfer*.

1.6. The first law of thermodynamics

The relationship between *heat* and *work* is stated by the first law of thermodynamics. As a preliminary manoeuvre, however, we must define what we mean by a *system*. Any part of the material world that can be marked off from its surroundings by real or imaginary boundaries may be regarded as a system. The system will then contain a definite and constant quantity or collection of matter. If the system is changing in shape then the boundaries must change with it so as to contain exactly the same material all the time. In other words, the boundaries must be flexible but impermeable. Since the system contains a definite quantity of matter, it will have for any particular state (defined by temperature, pressure, composition, etc.) a definite *internal energy*. For a discussion of the nature of internal energy the reader is referred to any of the standard textbooks on thermodynamics, for instance, Lewis and Randall, *Thermodynamics*, Chapter V[2].

If we now look at the system from the point of view of an observer who is *outside the system* we can say that, for a change from one state to another, the increase of internal energy is equal to the heat supplied *to* the system minus the work done *by* the system, i.e.

$$E - E_0 = Q - W \tag{1.6.1}$$

or, differentiating,

$$dE = dQ - dW, \tag{1.6.2}$$

but we must note that, whereas the internal energy is a function of the state (i.e. temperature, pressure, composition), the heat and work quantities depend on the particular process followed and not simply on the end-points. We have to be very careful, therefore, when using the differentials dQ and dW.

If we can give a satisfactory definition of the concepts of heat and work, we can regard the first law of thermodynamics almost as a definition of internal energy. Alternatively, if we can give a satisfactory statement of the concept of internal energy with the help of the kinetic theory, we can regard the first law of thermodynamics as providing a definition of what we mean by heat. For a fuller discussion see [2] and [3].

In writing equations (1.6.1) and (1.6.2) we have assumed the same units for heat and work and internal energy. It is customary to quote internal energy and heat quantities in *thermal units* (either British thermal units or centigrade heat units) and work in *mechanical units* (foot-pounds weight). The ratio between mechanical and thermal units is the *mechanical equivalent of heat, J*. Numerically 1 B.TH.U. = 778 ft.lb.wt., and 1 C.H.U. = 1400 ft.lb.wt. To save unnecessary use of symbols, however, we shall generally omit the factor J from our equations with the understanding that matters will be adjusted when the numbers are put in.

From a strictly thermodynamic point of view we can only observe a system from outside and can therefore only assess heat and work quantities by noting what is happening at the boundaries of the system. Heat quantities can only be measured directly by observing

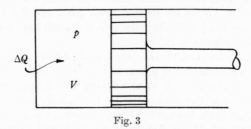

Fig. 3

the temperature gradient, and hence the local heat flux, across the boundary at every point. Nevertheless, it is sometimes helpful to picture what is happening to a system from the point of view of an observer who can move around inside, although this attitude is foreign to the ideas of thermodynamics.

Consider the case of a definite mass of gas expanding against a piston in an engine cylinder, as indicated in fig. 3. If we take the gas as the system then the walls of the cylinder and the piston constitute the boundaries. We will suppose that a restraining force is maintained on the piston rod so that external work is done when the piston moves. We will also suppose that the cylinder is jacketed with steam so that heat is being supplied to the system. The first law of thermodynamics states that, for a small movement of the piston,

$$\Delta E = \Delta Q - \Delta W. \tag{1.6.3}$$

If ΔW is numerically greater than ΔQ the system will be losing internal energy, i.e. ΔE will be negative, and vice versa.

If the expansion of the gas in the engine cylinder proceeds slowly, we can say that for a small volume change ΔV the work done by the system is $\Delta W = p\Delta V$, where p is the pressure of the gas. Note carefully that we can only make this substitution if we have conditions approximating to *mechanical equilibrium* between the system and its surroundings. If we now reduce the restraining force on the piston and allow the gas to expand rapidly, ΔW will be less than $p\Delta V$. In this case, part of the energy which was previously being transferred across the boundaries of the system in the form of mechanical work is now remaining within the system, and there is a dissipative process going on, associated with eddying and friction within the gas, which is helping to maintain the internal energy of the system. The external observer knows nothing about the details of the internal motion of the gas, but he can measure ΔW directly, and he therefore writes down the first law of thermodynamics in the form of equation (1.6.3) as before. He might refer to the difference between the maximum theoretical work $p\Delta V$ and the actual measured work ΔW as the friction work ΔW_f, but this term would not appear in his equation because it is not work that is transferred across the boundaries of the system.

An internal observer, on the other hand, knowing nothing about the restraining force on the piston rod and only able to measure the pressure, temperature and local velocity of the gas, would argue in either case that the work done by the system was $p\Delta V$. In the free expansion case, however, he would notice the internal motion of the gas and would conclude that due to the dissipative eddying process there was an internal release of frictional heat ΔQ_f. His application of the first law would therefore be

$$\Delta E = (\Delta Q + \Delta Q_f) - p\Delta V. \qquad (1.6.4)$$

Since the friction work and the friction heat are simply different interpretations of the same thing we conclude that

$$\Delta W_f = \Delta Q_f = p\Delta V - \Delta W, \qquad (1.6.5)$$

and equations (1.6.3) and (1.6.4) are therefore identical. Of the two, however, the external observer is making a stricter use of the methods of thermodynamics and the term friction work is generally preferable to the term friction heat.

There are other cases where it is sometimes convenient to use the idea of an internal release of heat, for example, the heating of a conductor carrying an electric current or the heat evolved in a chemical reaction. In the case of the conductor of resistance r carrying an electric current i it is customary to talk about the i^2r heating effect, although the strict application of the first law would be to say that electrical work was being done on the system at the rate i^2r by the passage of the current i against a potential difference ir.

In the case of a chemical reaction taking place at constant volume in a closed vessel we use the term *heat of reaction* to describe the change of internal energy ΔE under conditions of constant temperature. Each heat of reaction is given for a single temperature, but we do not mean that the temperature may not vary during the course of the reaction. What we really measure is the heat that must be transferred to or from the system in order to restore the temperature to its original value. If the system was thermally insulated so that no heat transfer could take place there would be no change of internal energy, but there would be a change of temperature of the system as a result of the reaction. A negative value for ΔE means that, when the original temperature has been restored, the products of the reaction have a lower internal energy and the reaction is therefore exothermic. A positive value for ΔE means that the reaction is endothermic. In chemical engineering we are more frequently concerned with chemical reactions which take place at *constant pressure*, and in these circumstances we use the term heat of reaction to describe the *change of enthalpy ΔH* under conditions of equal inlet and outlet temperatures.

Another example of an apparent internal release of heat occurs with substances which are undergoing radioactive decay or nuclear fission. In the case of these nuclear reactions we can again attribute the evolution of heat to a reduction in the internal energy of the system. Since mass is identified with energy, and since the energy release in nuclear reactions is sometimes very large, the change of internal energy may in certain cases be measurable as a change of mass of the system. This is equally true in principle of a chemical reaction, but since chemical energy changes are much smaller in magnitude the corresponding changes of mass are altogether insignificant.

1.7. Conduction, convection and radiation

The normal process by which heat is transferred within a solid is that of *conduction*. We have the experimental law that the heat flux at any point is equal to the *thermal conductivity k* of the material multiplied by the local temperature gradient

$$q = -k\frac{\mathrm{d}\theta}{\mathrm{d}r}. \tag{1.7.1}$$

If we attribute independent dimensions to heat (H) and to temperature (Θ), then the dimensions of thermal conductivity are given by $HL^{-1}T^{-1}\Theta^{-1}$, and numerical values are usually quoted in B.TH.U./ft.hr.°F. or C.H.U./ft.hr.°C. If, on the other hand, we identify heat dimensionally with mechanical energy, thermal conductivity will have dimensions $MLT^{-3}\Theta^{-1}$.

When we consider heat transfer within a fluid we again have the process of conduction, but this is usually supplemented by another process, that of *convection*, associated with the motion of the fluid. If the motion is maintained by an externally applied pressure difference as in the case of flow through a pipe, we use the term *forced convection*. If, on the other hand, the motion is due to density changes and the action of gravity, we use the term *free convection*. In turbulent flow we have another convective transfer mechanism at work due to the irregular eddying motion of the fluid.

We can account for the thermal conductivity of a gas in the same way that we can account for its viscosity with the help of the kinetic theory. In the one case we explain the transfer of energy and in the other the transfer of momentum due to the random motion of the molecules of the gas.

There is another heat-transfer process, that of *radiation*, which is quite independent of conduction and convection. Radiation is of particular importance in industrial processes when combustion is used as a source of heat. Since, however, radiation forms a separate subject on its own and has no direct connexion with fluid mechanics, we shall confine ourselves to *convective heat transfer*. The reader should consult [4] or [5] for a discussion of radiant heat transfer.

1.8. Diffusion and mass transfer

We have just noted that the phenomena of thermal conductivity and viscosity of a gas are explained by the kinetic theory of gases in terms of the transfer of energy and momentum by molecular motion. If we have a mixture of two different gases the same mechanism of molecular motion causes the transfer of mass by *diffusion*. Molecular diffusion also takes place in liquids although at much slower rates. In either case we can start with an approximate experimental law, similar to that for heat conduction, stating that the rate of mass transfer N of the diffusing component measured in mols per second per unit area is equal to the *diffusivity* $\mathscr{D}$ multiplied by the local concentration gradient of the diffusing component

$$N = -\mathscr{D}\,\frac{\mathrm{d}C}{\mathrm{d}r}. \tag{1.8.1}$$

The concentration C is expressed in mols per unit volume and the diffusivity $\mathscr{D}$ must therefore have dimensions L^2T^{-1}.

In addition to mass transfer by molecular diffusion we can have mass transfer by convection in a fluid which is in motion. There is, in fact, a general similarity between mass transfer and heat transfer by convective processes.

1.9. Mathematical note

There are two quite different uses of mathematics in engineering. The first, and more important use, is to enable numerical calculations to be made which can form the starting point of a design. The other function of mathematics in engineering is to enable certain general principles to be stated or derived in a clear and precise manner. In some instances the same piece of mathematical analysis can serve both purposes. Unfortunately, however, this is not generally true of fluid mechanics and heat transfer. The phenomena which we shall be dealing with are extremely complex, and when we attempt to formulate them in exact mathematical statements we generally find that we cannot easily handle the resulting equations. The procedure adopted in this book is to develop the main argument with the help of simple mathematics which is readily understood and suited to numerical illustration, and to summarize the more formal mathematical treatment of the subject in a series of appendices. At certain points, however, it is desirable to use some of the heavier mathematical equipment in the main text, but these sections are not essential to a first reading.

Vector notation is used occasionally. This is partly in order to save space but also because it is sometimes easier to visualize the meaning of the mathematics with the help of vector methods. A simple engineering guide to vector analysis is given in Appendix 1.

REFERENCES

(1)   JEANS. *Kinetic Theory of Gases* (Cambridge University Press).
(2)   LEWIS and RANDALL. *Thermodynamics* (McGraw-Hill).
(3)   KEENAN. *Thermodynamics* (Wiley).
(4)   FISHENDEN and SAUNDERS. *Heat Transfer* (Oxford).
(5)   MCADAMS. *Heat Transmission* (McGraw-Hill).

CHAPTER 2

FLUID FLOW

2.1. Continuity in fluid flow

Consider first of all the case of steady flow of a fluid. Fix attention on a particular point in the flow and picture a small closed curve drawn around this point in a plane perpendicular to the direction of flow (see fig. 4). The streamlines passing through this curve will form a *stream tube*, and if the curve is drawn small enough the velocity will be uniform over any cross-section of the stream tube. Since there is no flow across streamlines, conservation of mass requires that along the stream tube

$$\rho a u = \text{constant}, \tag{2.1.1}$$

where ρ is the density of the fluid, a is the cross-sectional area of the stream tube and u is the velocity of the fluid in the stream tube.

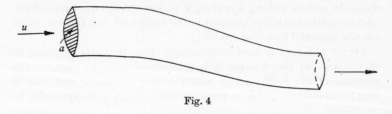

Fig. 4

This result can be stretched to cover the practical case of flow in a pipe of varying cross-section provided we are content to deal only in terms of the mean velocity and ignore the actual distribution of velocity across a section of the pipe.

2.2. The general equation of continuity†

Consider a fixed control surface S enclosing a volume V. The rate of accumulation of mass inside S must be equal to the rate of inflow across the control surface minus the rate of outflow. Referring to fig. 5, the rate of accumulation of mass inside the control surface is represented mathematically by $\dfrac{\partial}{\partial t} \oint_V \rho \, \mathrm{d}V$, and the *net* rate of

† This section may be omitted on a first reading. See Appendix 1 for a summary of vector notation and the methods of vector analysis.

outflow across the surface (i.e. outflow minus inflow) is given by the surface integral $\oint_S \rho\mathbf{u} \cdot \mathbf{n}\,dS$ or $\oint_S \rho u \cos\theta\,dS$, where θ is the angle between the velocity vector $\mathbf{u}$ and the normal to the surface $\mathbf{n}$ at any point on the surface as indicated in fig. 5. We can therefore write the equation

$$\frac{\partial}{\partial t}\oint_V \rho\,dV + \oint_S \rho u \cos\theta\,dS = 0. \qquad (2.2.1)$$

For the special case of *steady flow* this reduces to

$$\oint_S \rho u \cos\theta\,dS = 0. \qquad (2.2.2)$$

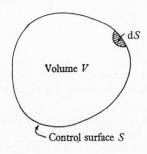

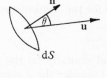

Fig. 5

By Gauss's theorem, the surface integral $\oint_S \rho\mathbf{u} \cdot \mathbf{n}\,dS$ is equal to the volume integral $\oint_V \operatorname{div}(\rho\mathbf{u})\,dV$. Equation (2.2.1) therefore becomes

$$\frac{\partial}{\partial t}\oint_V \rho\,dV + \oint_V \operatorname{div}(\rho\mathbf{u})\,dV = 0, \qquad (2.2.3)$$

and since this must be true however small the volume of the control surface, we can drop the integral sign and say that

$$\frac{\partial\rho}{\partial t} + \operatorname{div}(\rho\mathbf{u}) = 0, \qquad (2.2.4)$$

which is the general form of the equation of continuity. For *steady flow* this reduces to

$$\operatorname{div}(\rho\mathbf{u}) = 0, \qquad (2.2.5)$$

and for *steady incompressible flow* (i.e. with constant density ρ) it reduces further to

$$\operatorname{div}\mathbf{u} = 0 \tag{2.2.6}$$

or

$$\frac{\partial u}{\partial x} + \frac{\partial v}{\partial y} + \frac{\partial w}{\partial z} = 0.$$

2.3. The acceleration of a fluid element

For a fluid in motion, the velocity vector **u** at any given instant of time will be a function of position. In addition, unless the flow is steady, it will vary with time, i.e.

$$\mathbf{u} = f(\mathbf{r}, t) \tag{2.3.1}$$

or for the x component

$$u = f(x, y, z, t),$$

and similarly for the y and z components v and w.

Consider the x component u. The rate of change of u with time t *at a fixed point* is the 'local differential' $\partial u/\partial t$.

If we try to follow the motion of a *particular element of fluid*, however, we will want to know how its velocity changes as the element moves. Considering again the x component of velocity, the rate of change of the velocity component of the element is the 'individual' or 'total differential' Du/Dt *following the motion of the fluid*. We can derive an expression for this total differential as follows.

Suppose at time t the particle is at the point (x, y, z) at time $t + \mathrm{d}t$ it will be at $(x + u\mathrm{d}t, y + v\mathrm{d}t, z + w\mathrm{d}t)$, but $u = f(x, y, z, t)$.

$$\therefore \quad \mathrm{d}u = \frac{\partial u}{\partial t}\mathrm{d}t + \frac{\partial u}{\partial x}\mathrm{d}x + \frac{\partial u}{\partial y}\mathrm{d}y + \frac{\partial u}{\partial z}\mathrm{d}z$$

$$= \frac{\partial u}{\partial t}\mathrm{d}t + \frac{\partial u}{\partial x}u\mathrm{d}t + \frac{\partial u}{\partial y}v\mathrm{d}t + \frac{\partial u}{\partial z}w\mathrm{d}t$$

or

$$\frac{Du}{Dt} = \frac{\partial u}{\partial t} + u\frac{\partial u}{\partial x} + v\frac{\partial u}{\partial y} + w\frac{\partial u}{\partial z}; \tag{2.3.2}$$

the last three terms may be grouped together under the heading 'convective differential', i.e.

total differential = local differential + convective differential.

2.4. Differentiation following the motion of the fluid for any property H†

If H is any continuous function of $\mathbf{r}$ and t (scalar or vector)

$$H = f(\mathbf{r}, t), \tag{2.4.1}$$

and the total differential $\mathrm{D}H/\mathrm{D}t = \partial H/\partial t +$ convective differential. If at time t the particle is at the position $\mathbf{r}$, at time $t + \mathrm{d}t$ it will be

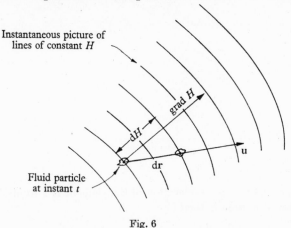

Instantaneous picture of
lines of constant H

grad H

dH

$\mathbf{u}$

dr

Fluid particle
at instant t

Fig. 6

at $\mathbf{r} + \mathrm{d}\mathbf{r}$ or $\mathbf{r} + \mathbf{u}\,\mathrm{d}t$. The change in H due to this displacement is given by the scalar product of $\mathrm{d}\mathbf{r}$ with grad H (see fig. 6), i.e.

$$\mathrm{d}H = \mathrm{d}\mathbf{r} \,.\, \nabla H = \mathbf{u} \,.\, \nabla H \,\mathrm{d}t,$$

and the convective differential $= \mathbf{u} \,.\, \nabla H$, therefore

$$\frac{\mathrm{D}H}{\mathrm{D}t} = \frac{\partial H}{\partial t} + \mathbf{u} \,.\, \nabla H. \tag{2.4.2}$$

If H happens to be a vector we can still say

$$\frac{\mathrm{D}H}{\mathrm{D}t} = \frac{\partial H}{\partial t} + (\mathbf{u} \,.\, \nabla)H,$$

where $(\mathbf{u} \,.\, \nabla)$ is interpreted as the operator $\left(u\dfrac{\partial}{\partial x} + v\dfrac{\partial}{\partial y} + w\dfrac{\partial}{\partial z} \right)$.

2.5. Bernoulli's equation

We can now derive the well-known Bernoulli equation which applies to the idealized case of the flow of an inviscid or frictionless fluid along a stream tube.

† This section may be omitted on a first reading.

Referring to fig. 7, let s be distance measured along the axis of the stream tube and let h be height measured vertically above a fixed datum level. Consider an element of fluid of instantaneous length ds and of mass $\rho a \, ds$ moving along the tube. For *steady flow*, by

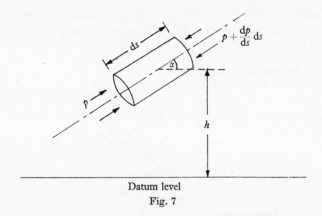

Datum level

Fig. 7

equation (2.3.2), the acceleration of the element is $u \, du/ds$, and the equation of motion is therefore

$$\rho a \, ds \, u \frac{du}{ds} = - a \frac{dp}{ds} ds - \rho a \, ds \, g \sin \alpha,$$

mass × acceleration   net force due   force due to
to pressure   gravity

i.e.
$$u \frac{du}{ds} = - \frac{1}{\rho} \frac{dp}{ds} - g \frac{dh}{ds}$$

or
$$\frac{1}{\rho} \frac{dp}{ds} + u \frac{du}{ds} + g \frac{dh}{ds} = 0. \qquad (2.5.1)$$

This is the Bernoulli equation in differential form without any restriction as to constant density. The equation can be integrated along the stream tube from section (1) to section (2) to give the more familiar form

$$\int_1^2 \frac{dp}{\rho} + \frac{u_2^2 - u_1^2}{2} + g \, (h_2 - h_1) = 0. \qquad (2.5.2)$$

For an *incompressible* fluid ρ is constant and the Bernoulli equation therefore becomes

$$\frac{p_1}{\rho} + \frac{u_1^2}{2} + gh_1 = \frac{p_2}{\rho} + \frac{u_2^2}{2} + gh_2. \qquad (2.5.3)$$

Note carefully that the dimensions here are energy per unit mass, i.e. foot-poundals/lb. or ft.^{2}/sec.^{2} When putting numbers in the equation the pressure must be measured in absolute units, i.e. poundals/sq.ft.

If we divide through equation (2.5.3) by g we get

$$\frac{p_1}{\rho g} + \frac{u_1^2}{2g} + h_1 = \frac{p_2}{\rho g} + \frac{u_2^2}{2g} + h_2. \qquad (2.5.4)$$

Each term now has the dimensions of length, and it is customary to describe the items in equation (2.5.4) as pressure head, velocity head and static head respectively. This is simply an extension of the well-known terminology of hydrostatics where pressures are quoted as so many feet of water or inches of mercury. Note once again that the pressure p in equation (2.5.4) must be measured in absolute units. There is a very common but dangerous habit of omitting the g from the first term with the understanding that pressure will be measured in practical units, i.e. in lb.wt./sq.ft. This is correct *numerically*, since g cancels out numerically with 32·2 (the ratio between the two units of force, pound weight and poundal), but it is quite incorrect *dimensionally*. The proper procedure, if we wish to have Bernoulli's equation in a form in which we can immediately substitute numerical values of the pressure in practical units, is to use the specific weight w in place of ρg. The equation then becomes

$$\frac{p_1}{w} + \frac{u_1^2}{2g} + h_1 = \frac{p_2}{w} + \frac{u_2^2}{2g} + h_2. \qquad (2.5.5)$$

The pressure may now be measured in lb.wt./sq.ft., and the specific weight w in lb.wt./cu.ft., and we are still correct dimensionally.

The Bernoulli equation may be expressed verbally by saying that, in the absence of measurable shearing stresses due to viscosity or fluid turbulence, the *total head* representing the sum of the pressure head, the velocity head and the static head *is constant along a streamline*. The conditions under which the same constant will apply to all streamlines in the flow are discussed in Appendix 2, which deals with the general equations of motion of a non-viscous fluid.

It might at first be thought that the Bernoulli equation would be of very limited application because of the restriction to frictionless flow. In fact, however, it has a surprisingly wide application particularly in the theory of flow-measuring devices, for instance, the venturi-tube, where we generally take care to avoid serious losses due to friction and turbulence. Other applications are discussed in Chapter 4.

2.6. Vortex motion

Consider an element of fluid moving in a circular path on a horizontal plane as indicated in the plan view of fig. 8. Let v be the tangential velocity of the element of fluid at radius r. The pressure gradient for radial equilibrium must be given by

$$\frac{\partial p}{\partial r} = \frac{\rho v^2}{r}, \qquad (2.6.1)$$

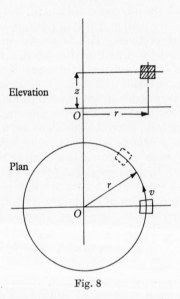

Fig. 8

and for vertical equilibrium

$$\frac{\partial p}{\partial z} = -\rho g, \qquad (2.6.2)$$

which is the ordinary hydrostatic condition.

In the special case of a *forced vortex* the angular velocity is constant, i.e. $v/r = \Omega = $ constant, and equation (2.6.1) becomes

$$\frac{\partial p}{\partial r} = \rho r \Omega^2. \qquad (2.6.3)$$

Integrating (2.6.2) and (2.6.3) therefore for the pressure distribution in a forced vortex

$$\frac{p - p_0}{\rho} = \tfrac{1}{2}\Omega^2 r^2 - g(z - z_0), \qquad (2.6.4)$$

and surfaces of constant pressure are paraboloids of revolution as indicated in fig. 9. A free surface is a special case of a surface of constant pressure and is therefore also a paraboloid of revolution.

A forced vortex is an extreme example of *rotational* flow, i.e. flow in which the fluid has *vorticity*. Vorticity is defined mathematically as curl **u**, where **u** is the velocity vector (see Appendix 3). The physical meaning of vorticity is best appreciated by picturing a small spherical element of fluid which is suddenly frozen solid. If the solid sphere is found to be rotating then the fluid has vorticity at the point considered. It is shown in Appendix 3 that the numerical value of the vorticity is twice the angular velocity of the sphere. In

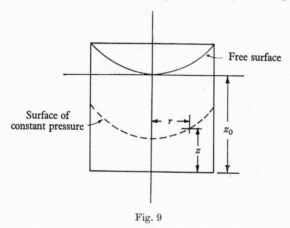

Fig. 9

the special case of the forced vortex the fluid is rotating as a solid body and every element must therefore have angular velocity Ω. The vorticity in a forced vortex is therefore constant and equal to 2Ω.

It follows from equation (2.6.4) that the Bernoulli total head for a forced vortex is given by

$$\frac{p}{\rho g} + \frac{v^2}{2g} + z = \frac{p_0}{\rho g} + \frac{\Omega^2 r^2}{g} + z_0, \tag{2.6.5}$$

i.e. the total head increases with radius. Flow of this type could neither originate nor survive on its own. Owing to the action of viscosity and fluid turbulence there would be a tendency for the mechanical energy or total head to be equalized throughout the flow. A forced vortex must, as its name implies, be created and maintained artificially by the action of an impeller or stirrer. Practical examples of a forced vortex are the flow in the impeller of a centrifugal pump operating at shut-off with the delivery valve closed, the centrifuge, and the central core of a stirred mixing vessel.

The conditions in a *free vortex* are quite different. For a free vortex the Bernoulli total head is uniform. By differentiating the Bernoulli equation with respect to radius, we have

$$\frac{1}{\rho}\frac{\partial p}{\partial r} + v\frac{\partial v}{\partial r} = 0 \qquad (2.6.6)$$

for constancy of total head along a radius. Substituting for $\partial p/\partial r$ from (2.6.1) we have

$$\frac{v}{r} + \frac{\partial v}{\partial r} = 0, \qquad (2.6.7)$$

i.e.

$$vr = \text{constant} = c. \qquad (2.6.8)$$

In other words, for a free vortex the angular momentum is constant. Substituting from (2.6.8) in (2.6.1) the radial pressure gradient for a free vortex is given by

$$\frac{\partial p}{\partial r} = \frac{\rho c^2}{r^3}. \qquad (2.6.9)$$

Integrating (2.6.2) and (2.6.9) therefore for the pressure distribution in a free vortex

$$\frac{p - p_0}{\rho} = g(z_0 - z) - \frac{c^2}{2r^2}, \qquad (2.6.10)$$

and the free surface is given by putting $p = p_0 = $ atmospheric pressure, i.e.

$$z = z_0 - \frac{c^2}{2gr^2}. \qquad (2.6.11)$$

It is shown in Appendix 3 that the vorticity is zero in a free vortex except at the origin. This may seem paradoxical at first sight, but a free vortex is in fact a special case of *irrotational motion*. A small spherical element of the fluid, suddenly frozen, would *not* be found to be spinning on its own axis.

The Bernoulli total head for a free vortex, which by definition is constant, is given by

$$\frac{p}{\rho g} + \frac{v^2}{2g} + z = \frac{p_0}{\rho g} + \frac{v^2}{2g} - \frac{c^2}{2gr^2} + z_0 = \frac{p_0}{\rho g} + z_0, \quad (2.6.12)$$

i.e. it is equal to the total head of the free surface at a great distance from the centre of the vortex. The form of the free surface is illustrated in fig. 10.

Practical examples of a free vortex are the whirlpool, and very approximately the outer portion of a stirred mixing vessel, the flow in the volute casing of a centrifugal pump, and the flow in a

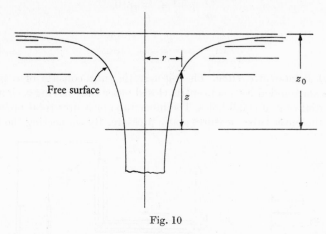

Fig. 10

cyclone dust collector. In the last three instances, however, friction at the wall has a modifying effect on the flow.

2.7. Flow measurement

It will be convenient to conclude this chapter with some notes on flow measurement. The following devices are all based on the principles of continuity and the Bernoulli equation.

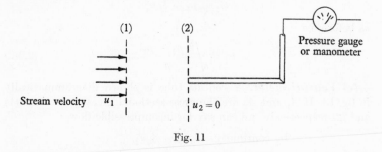

Fig. 11

(a) Pitot tube. A pitot tube is simply an open-ended tube pointed in the direction of the flow and connected to a pressure gauge or manometer, as indicated in fig. 11, and it measures the *total pressure*, i.e. the sum of the *static pressure* and the *dynamic* *pressure.* Applying Bernoulli from (1) to (2) for an incompressible fluid

$$\frac{p_1}{\rho} + \frac{u_1^2}{2} = \frac{p_2}{\rho} + 0,$$

i.e.

$$p_2 = p_1 + \tfrac{1}{2}\rho u_1^2. \tag{2.7.1}$$

$$\underset{\text{pressure}}{\text{total}} = \underset{\text{at section (1)}}{\text{static pressure}} + \underset{\text{pressure}}{\text{dynamic}}$$

(b) Pitot-static tube. The pitot-static tube consists of a pitot tube surrounded by a concentric closed outer tube having a circumferential ring of small holes. The inner tube measures total pressure and the outer tube measures static pressure. By connecting the two

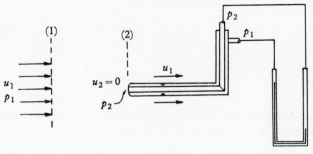

Fig. 12

tubes across a manometer as indicated in fig. 12 we can measure the *dynamic pressure* directly. If we know the density of the fluid we can calculate the velocity

$$p_2 = p_1 + \tfrac{1}{2}\rho u_1^2$$

as before, i.e.

$$u_1 = \sqrt{\frac{2(p_2 - p_1)}{\rho}}. \tag{2.7.2}$$

(c) Venturi-meter. A venturi tube is shown diagrammatically in fig. 13. If A_1 and A_2 are the cross-sectional areas at sections (1) and (2) respectively, we can say for incompressible flow

by continuity $A_1 u_1 = A_2 u_2,$

and by Bernoulli $\dfrac{p_1}{\rho} + \dfrac{u_1^2}{2} = \dfrac{p_2}{\rho} + \dfrac{u_2^2}{2},$

hence, $$u_1 = \sqrt{\frac{2(p_1 - p_2)}{\rho(A_1^2/A_2^2 - 1)}}, \tag{2.7.3}$$

or the theoretical discharge, assuming uniform velocity across the tube at either section, is given by

$$Q_{\text{theor.}} = A_1 u_1 = A_1 \sqrt{\frac{2(p_1 - p_2)}{\rho(A_1^2/A_2^2 - 1)}}. \tag{2.7.4}$$

In practice the discharge will be less than the theoretical value given by equation (2.7.4) owing to non-uniformity of the velocity distribution and to frictional effects. We therefore multiply the theoretical discharge by a *discharge coefficient C*, and we say that

$$Q_{\text{actual}} = CQ_{\text{theor.}},$$

i.e.
$$Q = CA_1 \sqrt{\frac{2(p_1 - p_2)}{\rho(A_1^2/A_2^2 - 1)}}. \tag{2.7.5}$$

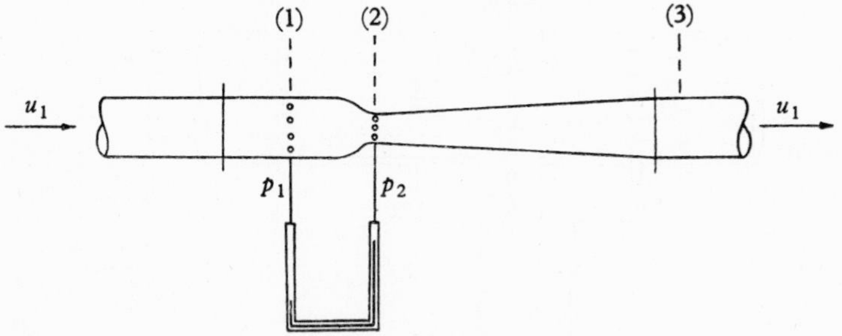

Fig. 13

The numerical value of the discharge coefficient will depend on the ratio A_1/A_2 and on the Reynolds number $u_2 d_2/\nu$ at the throat, but for a well-designed venturi-meter it should be in the range between 0·96 and 0·98. The significance of Reynolds number is explained in Chapter 5.

The shape of a venturi tube is important. It will be noted that there is a sharp contraction from section (1) to the throat at section (2) followed by a gradual expansion. The reason for this is that from (1) to (2) the fluid is being accelerated and the pressure is falling and there is no difficulty in achieving rapid acceleration of the flow in this manner. From (2) to (3), however, the fluid is being decelerated and the pressure is rising. If we try to achieve this pressure recovery too rapidly, the main stream of the fluid is liable to break away from the walls owing to a boundary-layer effect which will be explained in detail in Chapter 13. The expanding section of the tube is known as a *diffuser*. The more gradual the expansion of the diffuser the less will be the tendency for separation of flow to occur.

Ideally there should be complete recovery of pressure so that p_3 should be equal to p_1. In practice, however, even if there is no separation of flow in the diffuser, there will be some loss of pressure due to skin friction so that p_3 will inevitably be slightly less than p_1.

We have confined ourselves to *incompressible flow*. The theory of the venturi tube with compressible flow will be dealt with at a later stage.

(d) Orifice plate. The orifice plate is an alternative to the venturi-meter for measuring the flow in a pipe-line. The stream will emerge from the orifice, as shown in fig. 14, in the form of a submerged jet. Let A be the cross-sectional area of the pipe and let a be the area of

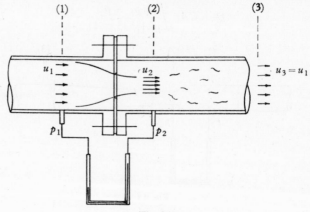

Fig. 14

the orifice. The jet converges to a vena-contracta just downstream from the orifice plate, and then breaks up into a region of violent eddying and turbulent flow. The ratio of the area of the jet at the vena-contracta to the area a of the orifice is known as the *coefficient of contraction*, and for a sharp-edged orifice plate of the type indicated in the diagram the numerical value of this coefficient is usually in the range from 0·62 to 0·70. The discharge Q is given by an expression similar to that for the venturi-meter

$$Q = CA \sqrt{\frac{2(p_1 - p_2)}{\rho(A^2/a^2 - 1)}}, \qquad (2.7.6)$$

but the *coefficient of discharge* C now includes both the effects of non-uniformity of velocity and the contraction of the jet. The numerical value of C will again depend on the ratio A/a and on the Reynolds number ud/ν, but is usually in the range from 0·61 to 0·62. It will be noted that the orifice plate is arranged with the sharp edge on the upstream side. The purpose of this is to ensure reproducible values for the coefficients of contraction and discharge and to avoid the necessity of separately calibrating every plate.

The orifice plate is particularly convenient, from a practical point of view, for flow measurement in pipe-lines. An orifice plate can easily be introduced at a flanged joint as suggested in fig. 14. It is not essential to have the pressure tappings in the pipe wall at sections (1) and (2) however; they can be located with very little error in the flanges themselves close on either side of the orifice plate.

There will be some partial recovery of pressure from section (2) to section (3), but the recovery will not be as good as that in the diffuser section of a venturi tube. The overall pressure drop in the pipe-line $p_1 - p_3$ will therefore be greater than that for a venturi-meter. An orifice plate is therefore more expensive as a flow-measuring device in running cost, but it is more convenient and cheaper than a venturi-meter in installation cost.

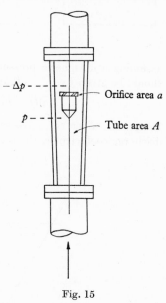

Orifice area a

Tube area A

Fig. 15

(e) Rotameter. All the devices mentioned so far depend on the measurement of a variable pressure difference. A rotameter, on the other hand, depends on the measurement of a variable flow area by means of the position of a float inside a tapered glass tube. The device is shown diagrammatically in fig. 15.

We can regard the annular passage between the float and the tapered tube as an orifice and apply the same equation as for the orifice plate. The pressure difference Δp, however, is almost constant and is fixed by the weight of the float. The flow will be given by

$$Q = CA \sqrt{\frac{2\Delta p}{\rho(A^2/a^2 - 1)}}. \qquad (2.7.7)$$

The ratio A/a is determined by the geometry of the instrument and will depend on the position of the float. The area A will also depend directly on the position of the float. By measuring the height of the float, therefore, the flow can be determined. In practice we do not use equation (2.7.7). It is simpler to calibrate each instrument and provide a scale of volume flow against float height.

(f) Sharp-crested weirs. The normal method of measuring flow in an open watercourse or channel is by means of a weir. A sharp-crested rectangular weir is shown in fig. 16. If we consider a strip of width dz at a height z above the crest of the weir, we can say that,

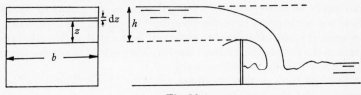

Fig. 16

assuming atmospheric pressure across the jet, ideally the velocity should be given by $u = \sqrt{\{2g(h-z)\}}$ and for the discharge

$$dQ = b\,dz\,\sqrt{\{2g(h-z)\}}.$$

Integrating over the cross-section of the weir and introducing a discharge coefficient c, the total discharge is given by

$$Q = \tfrac{2}{3}c\,\sqrt{(2g)}bh^{\frac{3}{2}}. \tag{2.7.8}$$

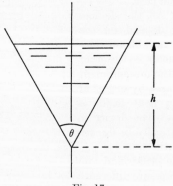

Fig. 17

For a triangular weir or V notch with cross-section as shown in fig. 17, a similar argument leads to the result

$$Q = \tfrac{8}{15}c\,\sqrt{(2g)}\tan\tfrac{1}{2}\theta\,h^{\frac{5}{2}}. \tag{2.7.9}$$

It will be apparent that, if the geometry of a weir is fixed, the only independent variable is the head h. If the head is constant a weir becomes a *flow-controlling device* rather than an instrument for measuring a variable flow.

(g) Broad-crested weir. The flow of a liquid over a broad-crested weir is shown diagrammatically in fig. 18. We assume that, where the flow becomes parallel, the velocity is uniform and the pressure distribution is hydrostatic. Applying the Bernoulli equation, the velocity u at the end of the weir will be given by

$$h = \lambda + \frac{u^2}{2g}$$

or

$$u = \sqrt{\{2g(h - \lambda)\}}.$$

The discharge

$$Q = b\sqrt{(2g)}\lambda(h - \lambda)^{\frac{1}{2}},$$

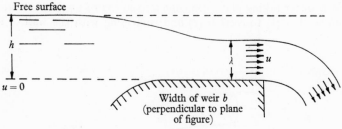

Fig. 18

for *maximum discharge*

$$dQ/d\lambda = 0 \quad \text{at constant } h,$$

i.e.

$$(h - \lambda)^{\frac{1}{2}} - \tfrac{1}{2}\lambda(h - \lambda)^{-\frac{1}{2}} = 0,$$

hence,

$$h = \tfrac{3}{2}\lambda \quad \text{or} \quad \lambda = \tfrac{2}{3}h,$$

and

$$Q_{\text{max.}} = \frac{2}{3\sqrt{3}}\sqrt{(2g)}bh^{\frac{3}{2}}. \tag{2.7.10}$$

The velocity u corresponding to this maximum rate of discharge is given by

$$u = \sqrt{(\tfrac{2}{3}gh)} = \sqrt{(\lambda g)}.$$

This is known as the *critical velocity* and is equal to the velocity of propagation of a surface wave. This critical value of the velocity in open channel flow is in some ways analogous to the sonic velocity in the compressible flow of a gas through a nozzle. Further discussion of open channel flow is deferred until Chapter 19.

CHAPTER 3

THE ENERGY AND MOMENTUM EQUATIONS

3.1. The energy equation for steady flow along a streamline

Consider again the case of the steady flow of a real fluid along a streamline or through a stream tube of small cross-sectional area as indicated in fig. 19. We will now make due allowance for the action of fluid friction and we will also include the possibility of heat transfer taking place. Imagine a fixed *control surface* extending from section (1) to section (2) and coinciding with the wall of the

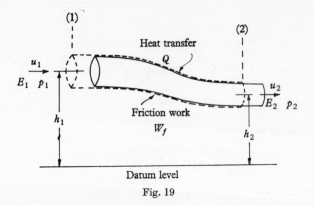

Fig. 19

stream tube between these points. The only flow across the control surface will be at sections (1) and (2). In fig. 19 the dotted outline represents the control surface which is fixed in space while the fluid flows through it. The fluid which happens to be inside the control surface at a particular instant will, after an interval of time dt, occupy a different position as represented by the full outline in fig. 19. In time dt the mass entering at section (1) $dm = \rho_1 a_1 u_1 dt$ and similarly the mass leaving at section (2) $dm = \rho_2 a_2 u_2 dt$. For steady flow these must be equal by the principle of continuity.

The fluid entering at section (1) is being pushed by the fluid behind it, and the fluid leaving at section (2) is pushing the fluid in front of it. The net work done against the pressure at sections (1) and (2) in time dt *by the fluid* which happens to be inside the control surface at the start of the interval considered is therefore given by

$$p_2 a_2 u_2 \, \mathrm{d}t - p_1 a_1 u_1 \, \mathrm{d}t = \left(\frac{p_2}{\rho_2} - \frac{p_1}{\rho_1} \right) \mathrm{d}m.$$

This work done against the fluid pressure may be called the *flow work*. For unit mass flow, the flow work $= p_2/\rho_2 - p_1/\rho_1$.

Owing to the existence of tangential shearing stresses work is also being done *by* the fluid inside the control surface at the wall of the stream tube *on* the fluid which surrounds it. This may be called the *friction work*. Let W_f be the friction work for unit mass flow through the stream tube.

We will assume that heat transfer is taking place within the fluid. Let Q be the net rate of heat supply across the control surface to the fluid inside per unit mass flow through the stream tube. For convenience we will use the same units both for heat and mechanical energy quantities.

The *total energy* per unit mass of the fluid at any point will be given by the sum of the intrinsic *internal energy* E, the *directed kinetic energy* $\frac{1}{2}u^2$, and the *potential energy* gh. The principle of conservation of energy applied to the flow system of fig. 19 states that, for unit mass flow of the fluid,

gain of total energy between sections (1) and (2)

    = heat received − work done by the fluid per unit mass flow,

i.e. $(E_2 - E_1) + \left(\dfrac{u_2^2}{2} - \dfrac{u_1^2}{2} \right) + g(h_2 - h_1) = Q - W_f - \left(\dfrac{p_2}{\rho_2} - \dfrac{p_1}{\rho_1} \right)$

or $E_1 + \dfrac{p_1}{\rho_1} + \dfrac{u_1^2}{2} + gh_1 + Q - W_f = E_2 + \dfrac{p_2}{\rho_2} + \dfrac{u_2^2}{2} + gh_2.$ (3.1.1)

Noting that the enthalpy H is defined by $H = E + p/\rho$, the last equation can be rewritten

$$H_1 + \tfrac{1}{2}u_1^2 + gh_1 + Q - W_f = H_2 + \tfrac{1}{2}u_2^2 + gh_2.$$ (3.1.2)

This is the general form of the energy equation for flow along a stream tube.

We can put equation (3.1.2) in differential form for a short length $\mathrm{d}s$ of the stream tube as indicated in fig. 20:

$$\mathrm{d}H + u \, \mathrm{d}u + g \, \mathrm{d}h = \mathrm{d}Q - \mathrm{d}W_f$$ (3.1.3)

or, since

$$H = E + pv, \quad \text{where} \quad v = \text{specific volume} = 1/\rho,$$

and therefore

$$\mathrm{d}H = \mathrm{d}E + p\,\mathrm{d}v + v\,\mathrm{d}p,$$

we can write $\mathrm{d}E + p\,\mathrm{d}v + v\,\mathrm{d}p + u\,\mathrm{d}u + g\,\mathrm{d}h = \mathrm{d}Q - \mathrm{d}W_f.$ (3.1.4)

If we now consider the particular infinitesimal element of fluid which happens to be inside the control surface of fig. 20 at a particular instant, we can imagine the element to be surrounded by a flexible boundary and we can treat it as a *system* in the strict thermodynamic sense of §1.6. Applying the first law of thermodynamics

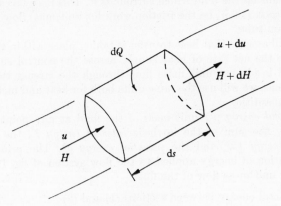

Fig. 20

to this element from the point of view of an *external observer moving with the fluid*,

gain of energy = heat received − work done by the element,

i.e. $$\mathrm{d}E = \mathrm{d}Q - p\,\mathrm{d}v.$$ (3.1.5)

Friction work does not appear in this equation because, although the observer views the element from outside, he is moving with the fluid and is therefore unable to detect any absolute motion. The observer can only detect the expansion work $p\,\mathrm{d}v$.

If we now subtract equation (3.1.5) from (3.1.4) we are left with

$$v\,\mathrm{d}p + u\,\mathrm{d}u + g\,\mathrm{d}h = -\,\mathrm{d}W_f.$$ (3.1.6)

This should be compared with the differential form of the Bernoulli equation (2.5.1). The left-hand side of (3.1.6) is the differential change in the Bernoulli energy. In §2.5 this was shown to be zero for frictionless flow. We now see that the mechanical or Bernoulli energy *decreases*, as one would expect, as a result of the action of friction and turbulence. The general energy equation (3.1.4) may be regarded as the sum of the mechanical energy equation (3.1.6) and the first law of thermodynamics as applied to a fluid element in equation (3.1.5).

3.2. The energy equation for steady flow in a pipe

Consider next the case of the steady flow of a real fluid through a pipe or any system with *fixed boundaries* as indicated in fig. 21. Take a control surface extending from section (1) to (2) and coinciding with the walls of the pipe or system between these sections.

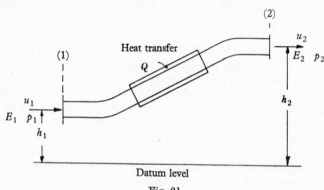

Fig. 21

The only work being done at the control surface by the fluid inside will now be the flow work $p_2/\rho_2 - p_1/\rho_1$ observable at sections (1) and (2). It is true that, owing to the action of fluid friction or viscosity, there are shear stresses at the walls. A real fluid, however, cannot slip at a solid boundary and therefore no shear work can be done at a fixed wall or surface. This is different from the case of the small stream tube considered in the preceding section because there we had a finite velocity at the edge of the system associated with the shear stresses. Similarly, if we have moving boundaries, for instance, in the case of flow through a row of turbine blades or through the impeller of a pump, the possibility of shear work again arises. Flow through a finite duct or system with moving boundaries will be dealt with in the next section; for the moment we will restrict ourselves to flow systems with *fixed boundaries*.

We will assume that heat transfer is taking place between the wall of the pipe and the fluid, and that Q is the net rate of heat supply per unit mass flow of fluid through the control surface. Applying the principle of the conservation of energy between sections (1) and (2) we arrive at the general energy equation in the form

$$E_1 + \frac{p_1}{\rho_1} + \frac{u_1^2}{2} + gh_1 + Q = E_2 + \frac{p_2}{\rho_2} + \frac{u_2^2}{2} + gh_2 \qquad (3.2.1)$$

or $$\qquad H_1 + \tfrac{1}{2}u_1^2 + gh_1 + Q = H_2 + \frac{u_2^2}{2} + gh_2. \qquad (3.2.2)$$

Note carefully that these equations apply only to flow through systems with solid boundaries which are fixed in space.

The differential form of (3.2.2) for a short length of pipe is

$$dH + u\,du + g\,dh = dQ \qquad (3.2.3)$$

or $$\qquad dE + p\,dv + v\,dp + u\,du + g\,dh = dQ. \qquad (3.2.4)$$

If we now picture the fluid which happens to be inside this short length at a particular instant as being surrounded by a flexible envelope we can apply the first law of thermodynamics. This is most easily done from the point of view of an observer who is able to move around inside the system, since the system extends across the entire section of the pipe. To such an observer the only work being done by the fluid would be the expansion work $p\,dv$. The apparent rate of heat supply, however, would be the heat dQ received from outside together with the heat dQ_f generated internally within the fluid as a result of friction and turbulence. The internal observer would therefore write the first law of thermodynamics in the following form (compare the argument in §1.6 leading to equation (1.6.4)):

$$dE = dQ + dQ_f - p\,dv. \qquad (3.2.5)$$

Subtracting equation (3.2.5) from (3.2.4) gives us the mechanical energy equation in the same form as equation (3.1.6)

$$v\,dp + u\,du + g\,dh = -dQ_f. \qquad (3.2.6)$$

We could equally well write dW_f in place of dQ_f, since friction work and friction heat are different interpretations of the same thing. The important point to bear in mind is that friction is an *internal* process.

3.3. The energy equation with shaft work

In the case of a fluid flowing past moving boundaries, as represented for instance by the blades of a turbine or compressor, the possibility of external work or shaft work arises. Consider the case of flow through a power plant or machine shown diagrammatically in fig. 22. In place of (3.2.2) the general energy equation now takes the form

$$H_1 + \tfrac{1}{2}u_1^2 + gh_1 + Q - W = H_2 + \tfrac{1}{2}u_2^2 + gh_2 \qquad (3.3.1)$$

or, if we neglect changes of level,

$$H_1 + \tfrac{1}{2}u_1^2 + Q - W = H_2 + \tfrac{1}{2}u_2^2. \qquad (3.3.2)$$

There are four important special cases of this general energy equation in engineering problems which should be mentioned.

(*a*) if there is no external work and no change of velocity, for instance, in flow through a heat exchanger, then the *heat transferred* = change of H;

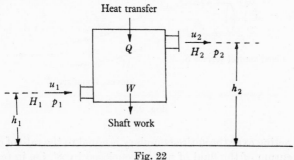

Fig. 22

(*b*) if there is no transfer of heat and no significant change of velocity, for instance, in flow through a compressor, then *external work* or *shaft work* = change of H;

(*c*) if there is no external work and no transfer of heat, for instance, in flow through a fixed nozzle, then the *gain of kinetic energy* of the fluid = change of H;

(*d*) if there is no external work, no transfer of heat, and no significant change of velocity, H must be constant, for example, in a throttling process.

3.4. The momentum equation for steady flow

The momentum principle expressed by Newton's second law in rigid body mechanics can only be applied directly to a *definite mass* of fluid. The extension of the principle to the flow system, for which we have already derived the continuity and energy equations, is almost obvious, but it is sufficiently important to justify a careful derivation.

In fig. 23 the dotted outline represents a *control surface S* which is fixed in space. In this particular diagram the control surface is shown as coinciding with a stream tube. This is convenient for visualizing the argument but it is not a necessary restriction. The argument will in fact apply to any imaginary fixed control surface that we like to choose. Consider next a flexible *moving boundary S′* enclosing a definite mass of fluid m and which always moves with the fluid so as to enclose the same mass m all the time. We will choose the moving boundary $S′$ so that it coincides with the fixed control surface S at time $t = t_0$. A moment later at $t = t_0 + dt$ it will have moved on to a new position as shown in the diagram. Applying Newton's law to the mass of fluid m enclosed by the boundary $S′$ we can say that the resultant force due to pressure and shearing

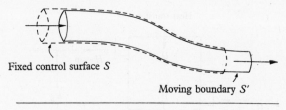

Fixed control surface S

Moving boundary $S′$

Fig. 23

stresses acting over the entire boundary is equal to the rate of change of momentum of the fluid of mass m enclosed by $S′$, i.e. in terms of the components in the x, y and z directions

$$F_x = \frac{dM_x}{dt}, \quad F_y = \frac{dM_y}{dt}, \quad F_z = \frac{dM_z}{dt},$$

or, in vector notation, $$\mathbf{F} = \frac{d\mathbf{M}}{dt}.$$

Provided the flow is steady,

the momentum of the fluid enclosed by the moving boundary $S′$ at time $t_0 + dt$

= momentum enclosed by the fixed surface S at time t_0

+ the outflow of momentum from the fixed surface S in time dt

− the inflow of momentum into the fixed surface S in time dt,

i.e.

the increase of momentum of the fluid enclosed by $S′$ in time dt

= the *net outflow of momentum* from the fixed surface S in time dt.

The momentum principle applied to a flow system with a fixed control surface, therefore, takes the following form:

$\mathbf{F}$ = net rate of outflow of momentum from the control surface,

$$(3.4.1)$$

where $\mathbf{F}$ is the resultant force due to pressure and shearing stresses acting over the entire surface.

A similar principle can be stated for *angular momentum* which is of importance in the theory of pumps and turbines. Taking a fixed control surface again, we can say that

$\mathbf{T}$ = net rate of outflow of angular momentum from the control surface, $(3.4.2)$

where $\mathbf{T}$ is the resultant torque due to pressure and shearing stresses acting over the surface.

Applications of these equations will be given in the following chapter.

CHAPTER 4

APPLICATIONS OF THE CONTINUITY, ENERGY AND MOMENTUM EQUATIONS

4.1. Sudden enlargement of a pipe

The problem is illustrated in fig. 24. We require to calculate the loss of head due to the sudden enlargement of a pipe. We will assume that the streamlines are parallel at section (2) and this will imply that the pressure p_2 is uniform across the pipe at this section.

By *continuity* with constant density

$$au_1 = au_2 = Au_3.$$

$$\therefore \quad u_2 = u_1 \quad \text{and} \quad u_3 = \frac{a}{A} u_1.$$

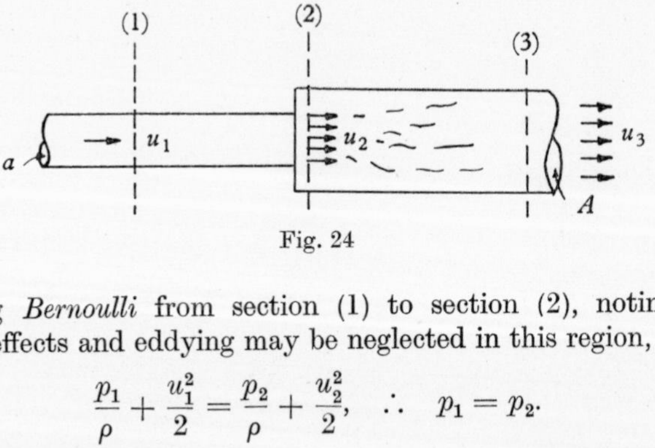

Fig. 24

Applying *Bernoulli* from section (1) to section (2), noting that friction effects and eddying may be neglected in this region,

$$\frac{p_1}{\rho} + \frac{u_1^2}{2} = \frac{p_2}{\rho} + \frac{u_2^2}{2}, \quad \therefore \quad p_1 = p_2.$$

The *momentum* equation may be applied to a cylindrical control surface extending from section (2) to (3). If we neglect skin friction at the wall of the pipe the resultant force acting over the entire control surface will be simply $(p_2 - p_3)A$, and this can be equated to the net rate of outflow of momentum, i.e.

$$(p_2 - p_3)A = \rho Au_3^2 - \rho au_2^2,$$

hence,

$$\frac{p_2 - p_3}{\rho} = u_3^2 - \frac{a}{A} u_2^2 = u_2^2 \left(\frac{a^2}{A^2} - \frac{a}{A} \right)$$

or

$$\frac{p_1 - p_3}{\rho} = u_1^2 \left(\frac{a^2}{A^2} - \frac{a}{A} \right). \tag{4.1.1}$$

38

Finally, from the *energy equation* the loss of mechanical energy per unit mass is given by

$$\left(\frac{p_1}{\rho} + \frac{u_1^2}{2}\right) - \left(\frac{p_3}{\rho} + \frac{u_3^2}{2}\right),$$

substituting for p_3 from (4.1.1) and for u_3 from the continuity equation, the loss of mechanical energy per unit mass is then given by

$$\frac{u_1^2}{2}\left(1 - \frac{a}{A}\right)^2,$$

i.e.

$$\text{loss of head} = \frac{u_1^2}{2g}\left(1 - \frac{a}{A}\right)^2, \tag{4.1.2}$$

If the area A is large compared with the area a the loss of head will approximate to $u_1^2/2g$, i.e. to the full inlet velocity head. It should be noted that we have only estimated the loss due to inertia effects associated with eddying and turbulence between sections (2) and (3). We are justified in neglecting skin friction in the momentum equation because we are only considering a short length of the pipe between sections (2) and (3).

As a numerical example suppose that water is flowing in a pipe at 20 ft./sec. and that there is a sudden enlargement of diameter to twice the original size so that $a/A = \frac{1}{4}$. The loss of head is found from (4.1.2) to be 3·5 ft. approximately. The difference in the static pressure $p_1 - p_3$ is found from (4.1.1) to be $- 1\cdot01$ p.s.i., i.e. if the static pressure at section (1) is atmospheric the static pressure at section (3) will be 1·01 p.s.i.g.

Note that pressure p_3 is greater than the pressure p_2, but the pressure increase is not as great as it would be if there was a gradual expansion or diffuser section connecting the small-diameter pipe to the larger one. The flow between sections (2) and (3) in this problem is essentially the same as that which occurs downstream from the vena contracta in the case of the orifice plate used for measuring flow rates in a pipe-line.

4.2. Reaction on a horizontal pipe bend

The problem is illustrated in fig. 25. We require to calculate the reaction on the pipe bend. p_1 and p_2 are expressed as gauge pressure in this instance. R_x and R_y in the figure are shown as the components of force exerted by the pipe bend *on* the fluid. We will take a control surface extending from section (1) to (2) and coinciding with the wall of the pipe between these sections:

$$Continuity: \qquad \rho a_1 u_1 = \rho a_2 u_2,$$

$$Bernoulli: \quad \frac{p_1}{\rho} + \frac{u_1^2}{2} = \frac{p_2}{\rho} + \frac{u_2^2}{2},$$

and by the *momentum* theorem the x component of the force acting over the control surface on the fluid inside is given by

$$F_x = p_1 a_1 - p_2 a_2 \cos \alpha - R_x = \rho Q(u_2 \cos \alpha - u_1),$$

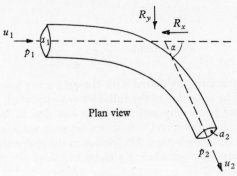

Fig. 25

and similarly for the y component

$$F_y = - p_2 a_2 \sin \alpha + R_y = \rho Q(u_2 \sin \alpha).$$

Hence, $\qquad R_x = p_1 a_1 - p_2 a_2 \cos \alpha + \rho Q(u_1 - u_2 \cos \alpha), \qquad (4.2.1)$

and $\qquad R_y = p_2 a_2 \sin \alpha + \rho Q u_2 \sin \alpha. \qquad (4.2.2)$

4.3. The jet pump or ejector

Fig. 26 shows the arrangement of a jet pump or ejector in which a supply of some fluid at high head is used to pump another flow from a low head to some intermediate head. In the following argument we will restrict ourselves to incompressible liquids:

Flow up to section (1):

For the driving jet: $\qquad h_j = \dfrac{p_1}{\rho g} + \dfrac{v_j^2}{2g}. \qquad (i)$

For the suction line: $\qquad h_s = \dfrac{p_1}{\rho g} + \dfrac{v_s^2}{2g}. \qquad (ii)$

It will be assumed that mixing of the two streams is complete at section (2).

From section (1) *to* (2):

    Continuity: $\rho_j v_j a_j + \rho_s v_s a_s = \rho_m u_2 A_2,$

or if $\rho_j = \rho_s = \rho_m,$ $u_2 = v_j \dfrac{a_j}{A_2} + v_s \dfrac{a_s}{A_2}.$ (iii)

    Momentum: $(p_1 - p_2)A_2 = \rho_m u_2^2 A_2 - \rho_j v_j^2 a_j - \rho_s v_s^2 a_s,$

or if $\rho_j = \rho_s = \rho_m,$ $\dfrac{p_1 - p_2}{\rho} = u_2^2 - v_j^2 \dfrac{a_j}{A_2} - v_s^2 \dfrac{a_s}{A_2}.$ (iv)

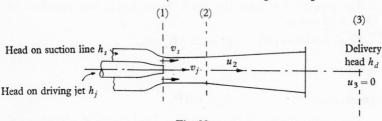

Fig. 26

From section (2) *to* (3):

    Energy: $\dfrac{p_2}{\rho} + \dfrac{u_2^2}{2} = \dfrac{p_3}{\rho} + 0 + L = gh_d + L,$

where L is the loss in the diffuser, i.e.

$$h_d = \frac{p_2}{\rho g} + \frac{u_2^2}{2g} - \frac{L}{g},$$ (v)

or if η_d is the efficiency of the diffuser, i.e. the ratio of the actual pressure recovery to the maximum recovery that is theoretically possible,

$$h_d = \frac{p_2}{\rho g} + \eta_d \frac{u_2^2}{2g}.$$ (v*a*)

If the three heads h_s, h_j and h_d, and the three areas a_s, a_j and A_2, are specified, we have five unknowns v_s, v_j, u_2, p_1 and p_2, and five equations for them which can be solved numerically. This will be clear from the following numerical example.

Water from a supply at 100 ft. head is used in a simple ejector to lift 10 cusecs of water from another source at a suction head of -5 ft. and to deliver the total quantity against a positive head. The nozzle area of the jet pipe is 0·05 sq.ft. and the annular area of the suction line at the throat section is 0·5 sq.ft. Determine the volume flow from the 100 ft. head supply and the delivery head of the ejector. The diffuser efficiency may be assumed to be 70 per cent.

If the volume flow in the suction line is 10 cusecs, $v_s = 20$ ft./sec. from equation (ii)

$$\frac{p_1}{\rho g} = h_s - \frac{v_s^2}{2g} = -11\cdot2,$$

∴ from (i)

$$\frac{v_j^2}{2g} = h_j - \frac{p_1}{\rho g} = 111\cdot2,$$

$$\therefore\quad v_j = 84\cdot5 \text{ ft./sec.}$$

The volume flow from the 100 ft. head supply will therefore be $Q_s = 4\cdot22$ cusecs.

From equation (iii) $u_2 = 25\cdot9$ ft./sec.,

and from (iv) $\dfrac{p_2}{\rho g} = -0\cdot5$ ft.;

hence, from (v) $h_d = 6\cdot6$ ft.

The simple theory of the ejector outlined above is adequate for the case of incompressible flow with liquids such as water. The problem becomes more complicated, however, if we have different fluids in the two lines which may be immiscible. Further difficulties arise if the driving fluid is a gas or vapour, for instance, air or steam, and the problem must then be treated as one of compressible flow. Both air and steam ejectors are widely used in the process industries, but their design is normally based on experiment rather than on theory.

4.4. The Pelton wheel

Water turbines can be classified either as impulse or as reaction machines. The characteristic feature of an *impulse* turbine is that no change of pressure occurs during passage of the water through the runner. The Pelton wheel is the normal form of impulse turbine, and this type is usually employed when there is a high head available and a relatively small volume flow of water. A discussion of the theory and design of water turbines is outside the field of this book. It will be convenient, however, to take one or two simple examples of calculations on water turbines for the purpose of illustrating some of the fundamental principles which have been stated in Chapters 2 and 3.

For a description of the main mechanical features of a Pelton wheel the reader is referred either to (1) at the end of this chapter

or to a turbine manufacturer's catalogue. It will be sufficient for
the purposes of the present example to say that the *runner* of the
turbine consists of a disk mounted on the main shaft and carrying
a number of double spoon-shaped *buckets* around its periphery. The
water is discharged from the nozzle in the form of a free jet directed
tangentially at the runner. Each bucket has a sharp splitting edge
at its centre-line and the jet therefore divides into two after impact

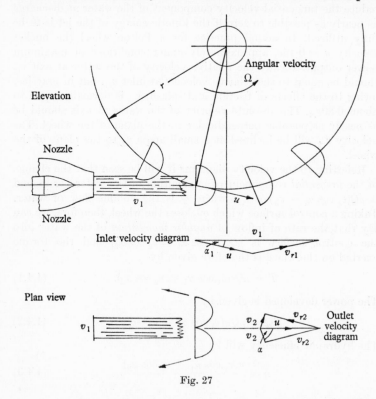

Fig. 27

and the water is turned outwards from the runner for the purposes
of discharge with a small component of velocity parallel to the
turbine shaft. Control is effected by means of a needle or spear-
shaped valve forming part of the nozzle. The valve is usually
operated by a governor acting through a servo-mechanism. Pelton
wheels are usually designed with either one or two nozzles associated
with a single runner.

Velocity diagrams at inlet to and outlet from the runner are
shown in fig. 27.

The jet velocity v_1 will be given by $v_1 = c_v \sqrt{(2gh)}$, where h is the effective head at the turbine nozzle and c_v is the velocity coefficient for the nozzle. The numerical value of c_v should be in the neighbourhood of 0·98 or 0·99 for a well-designed machine. For operation at maximum power the peripheral velocity of the buckets will be slightly less than half the jet velocity. This is a characteristic feature of any impulse machine and is a direct consequence of the need to reduce the tangential velocity component of the water at discharge as nearly as possible to zero if the kinetic energy of the jet is to be fully utilized. In actual practice for a Pelton wheel the bucket velocity $u = 0·46v_1$ approximately under conditions of maximum power output. Ideally the relative velocity of the water at exit v_{r2} should be equal to the relative velocity at inlet v_{r1}, but in practice, owing to the effects of friction and splash, v_{r2} is normally equal to about $0·80v_{r1}$. The absolute velocity of the water at exit should be as nearly as possible perpendicular to the plane of the wheel. The velocity v_{r2} will be inclined at a small angle β_2 to the plane of the wheel.

Referring to the velocity diagrams it will be seen that the change of the *tangential* component of momentum of the water per second is $\rho Q(v_1 \cos \alpha_1 - v_2 \cos \alpha_2)$, where Q is the volume flow of water. Taking a control surface which encloses the wheel, therefore, we can say that the rate of inflow of *angular momentum* of the water into the control surface is $\rho Q r(v_1 \cos \alpha_1 - v_2 \cos \alpha_2)$ and the torque exerted on the wheel is therefore given by

$$T = \rho Q r(v_1 \cos \alpha_1 - v_2 \cos \alpha_2). \tag{4.4.1}$$

The power developed is given by

$$T\Omega = \rho Q u(v_1 \cos \alpha_1 - v_2 \cos \alpha_2). \tag{4.4.2}$$

The hydraulic efficiency will be

$$\eta = \frac{u(v_1 \cos \alpha_1 - v_2 \cos \alpha_2)}{gh}. \tag{4.4.3}$$

As a numerical example let us calculate the maximum power output and some of the leading dimensions of a single jet Pelton wheel to operate under a head of 1000 ft. and to drive a generator at 300 r.p.m. Taking a velocity coefficient for the nozzle of 0·985 we find the jet velocity

$$v_1 = c_v \sqrt{(2gh)} = 250 \text{ ft./sec.}$$

Assuming that $u = 0·46\,v_1$ for maximum power, $u = 115$ ft./sec. but $u = \pi N D$, where the rotational speed $N = 300$ r.p.m. or 5 r.p.s., hence the pitch circle diameter of the wheel $D = 7·32$ ft. The flow of water and therefore the power output will depend on the nozzle area. It is usual for the jet diameter to be limited to about a twelfth of the wheel diameter. With these proportions the jet diameter will be 7·32 in. and the volume flow will be 73 cusecs. From the geometry of the velocity diagrams in fig. 27, taking typical values of $\alpha_1 = 10°$ and $\beta_2 = 15°$, we find $v_1 \cos \alpha_1 = 246$ and $v_2 \cos \alpha_2 = 8$, so that $(v_1 \cos \alpha_1 - v_2 \cos \alpha_2) = 238$ ft./sec. Hence from (4.4.2) the maximum power developed is given by

$$T\Omega = \frac{62·5 \times 73 \times 115 \times 238}{32·2 \times 550} = 7050 \text{ h.p.}$$

4.5. Reaction turbines

There are two types of reaction turbine commonly used in hydro-electric power plants, the radial flow or Francis turbine, and the axial or propellor turbine. Axial flow turbines with adjustable blades are known as Kaplan turbines. The axial flow type is suited for installations employing low heads and a large volume flow of water. The radial flow type, on the other hand, covers a wide range of head and flow intermediate between the axial flow turbine at one extreme and the Pelton wheel at the other.

In this example the momentum and energy equations will be applied to the flow of water through the runner of a radial flow reaction turbine. The basic mechanics is the same, however, whether we are dealing with radial or axial flow. Actually the flow in a Francis runner is not entirely radial. The water enters in a radial direction, but in its passage through the runner it is usually turned through 90° into an axial direction for discharge.

In fig. 28 velocity diagrams are shown at inlet to and outlet from the runner for the case of inward radial flow. The water enters the runner, after passing through fixed or movable guide vanes, with *absolute velocity* v_1. It is convenient to resolve this velocity into two components, a tangential component or *velocity of whirl* w_1, and a radial component or *velocity of flow* f_1. The peripheral *velocity of the runner* is denoted by u_1 and the *relative velocity* of the water to the runner by v_{r1}. A similar notation is used for the outlet velocity diagram.

The guide vanes are designed so that the relative velocity v_{r1} is tangential to the blades at inlet when the turbine is running at the design speed. The relative velocity v_{r2} will also be approximately tangential to the blades at outlet. In a reaction turbine v_1 will be less than $\sqrt{(2gh)}$, i.e. the full velocity head is not developed during passage of the water through the guide vanes, but pressure changes will occur as the water flows through the runner. The turbine must run full and the casing will be under pressure.

The *continuity equation* in conjunction with the geometry of the turbine will determine the flow velocity f_2 in relation to the flow velocity f_1.

The *angular momentum equation* enables us to calculate the shaft work in terms of the velocities. Taking a control surface which just encloses the runner, we can say that the torque exerted *by*

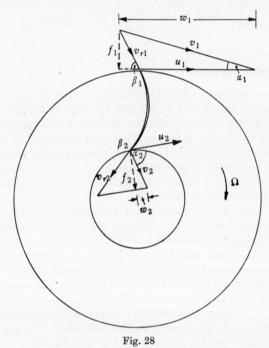

Fig. 28

the water *on* the runner is equal to the net rate of inflow of angular momentum into the control surface, i.e.

$$T = \rho Q(r_1 v_1 \cos \alpha_1 - r_2 v_2 \cos \alpha_2)$$

or
$$T = \rho Q(r_1 w_1 - r_2 w_2), \qquad (4.5.1)$$

and the power developed will be given by

$$T\Omega = \rho Q(w_1 u_1 - w_2 u_2), \qquad (4.5.2)$$

the *shaft work* per pound flow will therefore be

$$W = w_1 u_1 - w_2 u_2, \qquad (4.5.3)$$

expressed in absolute units, i.e. foot poundals/lb.

The mechanical *energy equation* can now be written down for a complete turbine installation such as that illustrated diagrammatically in fig. 29. If H is the gross head available, and if h_f is the friction loss in the penstock, etc., up to section (1), then neglecting losses in the runner itself, we can say

$$g(H - h_f) = \frac{p_1}{\rho} + \frac{v_1^2}{2} = \frac{p_2}{\rho} + \frac{v_2^2}{2} + W, \qquad (4.5.4)$$

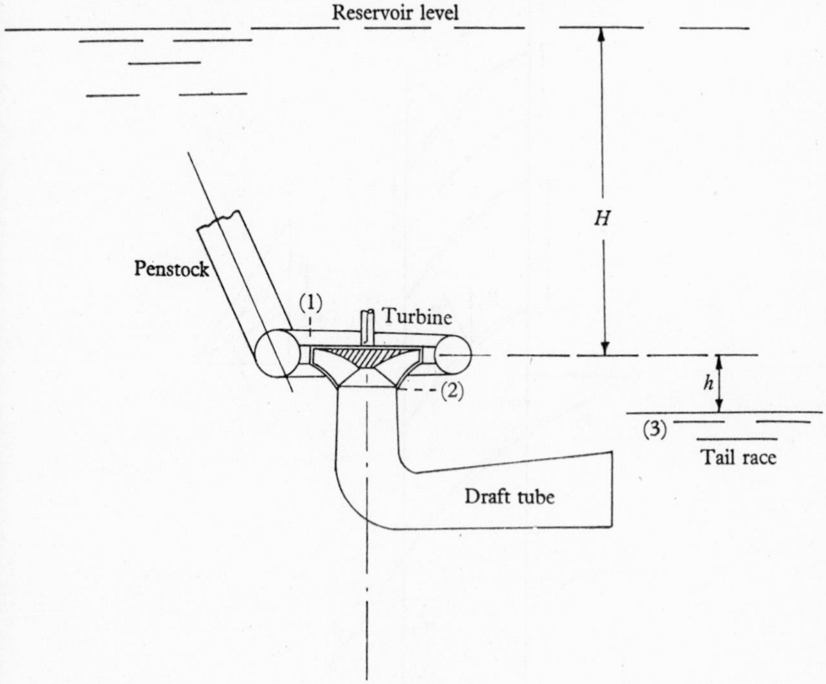

Fig. 29

where W is given by equation (4.5.3). For flow in the draft tube, assuming a loss of head h_d between (2) and (3)

$$\frac{p_2}{\rho} + \frac{v_2^2}{2} + g(h - h_d) = \frac{v_3^2}{2}, \qquad (4.5.5)$$

since $p_3 = 0$ gauge.

4.6. Flow past a cascade of vanes

The problem is illustrated in fig. 30. A cascade of vanes is used to deflect a stream of fluid. The absolute velocity of the fluid at entry to the vanes is denoted by v_1, and this may be resolved into an axial or flow component f_1 and a tangential component w_1. A similar notation is used for the outlet side.

By the *continuity equation* $f_1 = f_2 = f$, assuming constant flow area and an incompressible fluid, i.e. we have a constant flow velocity f.

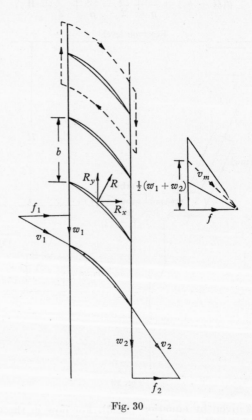

Fig. 30

Applying the *Bernoulli equation*, neglecting skin friction,

$$\frac{p_1}{\rho} + \frac{v_1^2}{2} = \frac{p_2}{\rho} + \frac{v_2^2}{2},$$

therefore

$$p_1 - p_2 = \tfrac{1}{2}\rho(v_2^2 - v_1^2) = \tfrac{1}{2}\rho(w_2^2 - w_1^2). \tag{4.6.1}$$

Taking a control surface, such as that shown by the dotted outline in the figure, enclosing one blade, and taking unit length perpendicular to the plane of the diagram, the *momentum* theorem may be applied as follows:

for the x direction:

$$(p_1 - p_2)b - R_x = 0; \tag{4.6.2}$$

for the y direction:

$$R_y = \rho f b(w_2 - w_1). \tag{4.6.3}$$

From (4.6.1) and (4.6.2), eliminating $p_1 - p_2$,

$$R_x = \tfrac{1}{2}\rho(w_1 + w_2)b(w_2 - w_1). \tag{4.6.4}$$

From (4.6.3) and (4.6.4), the resultant force R can be written

$$R = \rho v_m b(w_2 - w_1), \tag{4.6.5}$$

where v_m is the mean velocity vector, having components f and $\tfrac{1}{2}(w_1 + w_2)$, as indicated in fig. 30. Evidently the force R is perpendicular to the velocity vector v_m.

The quantity $b(w_2 - w_1)$ is the *circulation* round one of the blades. The circulation K is defined as the line integral of the velocity component taken round a closed circuit in the fluid (see Appendix 3). If we take a circuit, such as that shown by the dotted line in fig. 30, enclosing one of the blades and lying in the plane of the diagram, it will be seen that the contribution of the two curved portions cancels out exactly and we are left with

$$K = b(w_2 - w_1), \tag{4.6.6}$$

the evaluation being carried out in a clockwise direction.

We therefore arrive at the result that

$$R = \rho v_m K, \tag{4.6.7}$$

where R is the force per unit length of blade. This is a special case of a general theorem in aerodynamics which states that for an aerofoil of infinite span the lift force per unit length is

$$L = \rho U K,$$

where U is the velocity of the stream and K is the circulation.

4.7. Flow of a gas through a nozzle

We can now consider the case of the *compressible flow* of a perfect gas through a tube or nozzle such as that shown in fig. 31. We will restrict ourselves to one-dimensional flow, i.e. we will assume that the velocity is uniform over any cross-section of the nozzle.

The *continuity equation* states that the mass flow is constant, i.e.

$$m = \rho A u.$$

The *energy equation* can be applied in the form of (3.2.2), provided the nozzle is fixed and there is no external shaft work involved. Neglecting changes of level and assuming that there is no heat transfer, the equation reduces to the simple form

$$H_1 + \tfrac{1}{2}u_1^2 = H_2 + \tfrac{1}{2}u_2^2,$$

or if the flow originates from a reservoir where the velocity is zero and where the pressure, temperature and enthalpy are p_0, T_0 and H_0 respectively, we can say that

$$H_0 = H + \tfrac{1}{2}u^2. \qquad (4.7.1)$$

It is always convenient, when dealing with compressible flow in a nozzle or channel, to picture the stream originating from a reservoir in this way whether or not such a reservoir does in fact exist.

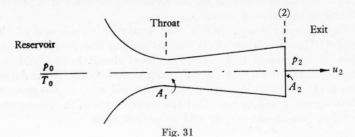

Fig. 31

For a perfect gas

$$H = c_p T = \frac{\gamma}{\gamma - 1} \frac{p}{\rho} = \frac{a^2}{\gamma - 1},$$

where c_p is the specific heat at constant pressure, c_v is the specific heat at constant volume, $\gamma = \dfrac{c_p}{c_v}$ and $a = \sqrt{(\gamma R T)}$. It can be shown that a is equal to the local speed of sound in the gas at temperature T. R is the gas constant defined in the ordinary engineering sense by the perfect gas law $p/\rho = RT$.

Equation (4.7.1) may therefore be expressed in any of the following forms, so long as we are dealing with a perfect gas:

$$c_p T_0 = c_p T + \tfrac{1}{2}u^2, \qquad (4.7.2)$$

$$\frac{\gamma}{\gamma - 1} \frac{p_0}{\rho_0} = \frac{\gamma}{\gamma - 1} \frac{p}{\rho} + \frac{u^2}{2}, \qquad (4.7.3)$$

$$\frac{a_0^2}{\gamma - 1} = \frac{a^2}{\gamma - 1} + \frac{u^2}{2}, \qquad (4.7.4)$$

or
$$\frac{T_0}{T} = \frac{a_0^2}{a^2} = 1 + \frac{\gamma - 1}{2} M^2, \tag{4.7.5}$$

where the Mach number M is defined by $M = u/a$. It is particularly important to appreciate that the above equations (4.7.1)–(4.7.5) are true even when frictional or other losses are involved. We have not yet assumed isentropic or frictionless flow, although we shall do so in a moment. Up to this point we have only assumed that the fluid is a perfect gas and that there is no external work and no heat transfer.

If the restriction to frictionless *isentropic flow* is now made we will have the additional equation

$$\frac{p}{p_0} = \left(\frac{\rho}{\rho_0}\right)^\gamma = \left(\frac{T}{T_0}\right)^{\gamma/(\gamma-1)}, \tag{4.7.6}$$

and it can also be shown that, for constant entropy, $a^2 = \mathrm{d}p/\mathrm{d}\rho$. We can now use the Bernoulli equation in differential form, which is valid for frictionless compressible flow,

$$\frac{1}{\rho}\,\mathrm{d}p + u\,\mathrm{d}u = 0$$

or
$$a^2\frac{\mathrm{d}\rho}{\rho} + u\,\mathrm{d}u = 0, \tag{4.7.7}$$

and from the continuity equation

$$\frac{\mathrm{d}\rho}{\rho} + \frac{\mathrm{d}u}{u} + \frac{\mathrm{d}A}{A} = 0. \tag{4.7.8}$$

Eliminating $\mathrm{d}\rho/\rho$ between (4.7.7) and (4.7.8) we have

$$\left(1 - \frac{u^2}{a^2}\right)\frac{\mathrm{d}u}{u} + \frac{\mathrm{d}A}{A} = 0$$

or
$$(1 - M^2)\frac{\mathrm{d}u}{u} = -\frac{\mathrm{d}A}{A}. \tag{4.7.9}$$

If $M < 1$ u increases as A decreases and vice versa, but if $M > 1$ u increases as A increases and we can only have $M = 1$ at a section of minimum area. Putting $M = 1$ in the energy equation (4.7.5) gives $T_0/T^* = \frac{1}{2}(\gamma + 1)$, and the corresponding pressure p^* at the point where $M = 1$ is given by

$$\frac{p^*}{p_0} = \left(\frac{2}{\gamma + 1}\right)^{\gamma/(\gamma-1)}, \tag{4.7.10}$$

and this is known as the *critical pressure ratio*.

It should be appreciated that the Mach number at the throat section is not necessarily equal to 1. It is perfectly possible to have compressible subsonic flow throughout the nozzle. If, however, the back pressure is low enough for the critical pressure to be reached at the throat then the nozzle is said to be *choked*, the Mach number will be 1 at the throat, and no further reduction of the back pressure will affect the mass flow.

If the nozzle is choked the following results apply for the temperature, density, and velocity at the throat:

$$\frac{T^*}{T_0} = \frac{2}{\gamma + 1}, \tag{4.7.11}$$

$$\frac{\rho^*}{\rho_0} = \left(\frac{2}{\gamma + 1}\right)^{1/(\gamma - 1)} \tag{4.7.12}$$

$$u^* = a^* = \sqrt{(\gamma R T^*)} = \sqrt{\left(\frac{2\gamma}{\gamma + 1} R T_0\right)}, \tag{4.7.13}$$

and the mass flow will be given by

$$m = \rho^* u^* A_t = \left(\frac{2}{\gamma + 1}\right)^{1/(\gamma - 1)} \rho_0 \sqrt{\left(\frac{2\gamma}{\gamma + 1} R T_0\right)} A_t$$

or $\qquad m = \left(\frac{2}{\gamma + 1}\right)^{1/(\gamma - 1)} \cdot \sqrt{\left(\frac{2\gamma}{(\gamma + 1)} p_0 \rho_0\right)} A_t. \tag{4.7.14}$

For *air* $\gamma = 1\cdot4$ and equation (4.7.14) reduces to

$$m = 0\cdot684\sqrt{(p_0 \rho_0)} A_t.$$

For *superheated steam* we can take $\gamma = 1\cdot3$, in which case

$$m = 0\cdot668\sqrt{(p_0 \rho_0)} A_t.$$

Note that absolute units must be used in these numerical equations, i.e. the pressure p_0 must be expressed in poundals/sq.ft., density ρ_0 in lb./cu.ft., area A_t in sq.ft., and the mass flow m will then be given in lb./sec.

A fuller account of compressible flow in nozzles, including the nature of the flow downstream from the throat, will be given in Chapter 18.

4.8. Normal shock wave

We can now investigate another feature of compressible flow while still employing only the simple tools represented by the continuity, energy and momentum equations. Suppose that we have a compressible gas moving at high velocity in a parallel stream as indicated in fig. 32. Is it possible to have a discontinuity in the pressure, i.e. is it possible for the pressure p_1 to change suddenly to another value p_2 with corresponding changes in the velocity and density? The question can be answered by writing down the conditions that must be fulfilled in order to satisfy the continuity, energy and momentum equations.

Continuity:
$$\rho_1 u_1 = \rho_2 u_2. \tag{i}$$

Energy, from (4.7.4) for the case of a perfect gas:
$$\frac{a_1^2}{\gamma - 1} + \frac{u_1^2}{2} = \frac{a_2^2}{\gamma - 1} + \frac{u_2^2}{2}. \tag{ii}$$

Momentum:
$$p_1 + \rho_1 u_1^2 = p_2 + \rho_2 u_2^2. \tag{iii}$$

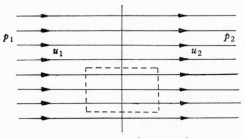

Fig. 32

Dividing equation (ii) by $\frac{1}{2} u_2^2$ and multiplying by $(\gamma - 1)$, we get
$$\left[\frac{2}{M_1^2} + (\gamma - 1) \right] \frac{u_1^2}{u_2^2} = \left[\frac{2}{M_2^2} + (\gamma - 1) \right], \tag{ii$a$}$$

and from (iii) noting that $a^2 = \gamma p / \rho$,
$$\rho_1 (a_1^2 + \gamma u_1^2) = \rho_2 (a_2^2 + \gamma u_2^2)$$
or
$$\rho_1 u_1^2 \left(\frac{1}{M_1^2} + \gamma \right) = \rho_2 u_2^2 \left(\frac{1}{M_2^2} + \gamma \right),$$

and making use of the continuity equation (i)
$$\frac{u_1}{u_2} \left(\frac{2}{M_1^2} + 2\gamma \right) = \left(\frac{2}{M_2^2} + 2\gamma \right). \tag{iii$a$}$$

Subtract (iii$a$) from (ii$a$)
$$\left[\frac{2}{M_1^2} + (\gamma - 1) \right] \left(\frac{u_1^2}{u_2^2} - \frac{u_1}{u_2} \right) - (\gamma + 1) \left(\frac{u_1}{u_2} - 1 \right) = 0,$$
i.e.
$$\left(\frac{u_1}{u_2} - 1 \right) \left\{ \left[\frac{2}{M_1^2} + (\gamma - 1) \right] \frac{u_1}{u_2} - (\gamma + 1) \right\} = 0. \tag{iv}$$

There are two possibilities. Either $(u_1/u_2 - 1) = 0$, i.e. $u_1 = u_2$ and nothing happens, or else

$$\frac{u_1}{u_2}\left[\frac{2}{M_1^2} + (\gamma - 1)\right] - (\gamma + 1) = 0, \tag{v}$$

which represents a normal shock wave in the flow.

If there is a shock wave, the following results can be derived from equations (i), (ii), (iii), and (v):

$$\frac{u_2}{u_1} = \frac{(\gamma - 1)M_1^2 + 2}{(\gamma + 1)M_1^2}, \tag{4.8.1}$$

$$M_2^2 = \frac{(\gamma - 1)M_1^2 + 2}{2\gamma M_1^2 - (\gamma - 1)}, \tag{4.8.2}$$

$$\frac{p_2}{p_1} = \frac{2\gamma M_1^2 - (\gamma - 1)}{(\gamma + 1)}, \tag{4.8.3}$$

and
$$\frac{T_2}{T_1} = \frac{(\gamma - 1)M_1^2 + 2}{(\gamma - 1)M_2^2 + 2}. \tag{4.8.4}$$

Anticipating the results of §18.3, it can be said that flow is only possible through a shock wave *from a lower to a higher pressure* and *from a supersonic to a subsonic velocity*. This is associated with the fact that, in this direction, the sudden change of state of the gas involves an *increase of entropy*, and the change is therefore thermodynamically possible in an isolated stream of fluid. A discontinuity leading from a higher to a lower pressure and from a subsonic to a supersonic velocity, on the other hand, would involve a decrease of entropy, and a spontaneous change in this direction would therefore contravene the second law of thermodynamics. It is important to appreciate the fact that although energy is conserved in the passage of the gas through the shock wave, there is an increase of entropy in the process and the change is therefore irreversible in the thermodynamic sense.

As a numerical example of the magnitude of the changes which can occur in a normal shock wave the following figures are readily evaluated for the case of air with $\gamma = 1.4$:

M_1	M_2	u_2/u_1	p_2/p_1	T_2/T_1
1·1	0·91	0·86	1·25	1·07
1·5	0·70	0·54	2·46	1·32
2·0	0·58	0·38	4·50	1·69

REFERENCES

(1) GIBSON. *Hydraulics and its Applications* (Constable).
(2) PRANDTL. *The Essentials of Fluid Dynamics* (Blackie).

CHAPTER 5

DIMENSIONAL ANALYSIS APPLIED TO FLUID MECHANICS

5.1. Statement of the principle

All mechanical quantities can be expressed in terms of a limited number of *fundamental dimensions*. If we exclude thermal and electrical phenomena for the time being, we can say that only three fundamental dimensions are necessary, mass, length and time, for which we shall use the symbols M, L, T. Other mechanical quantities, such as velocity, acceleration, force, energy, momentum, etc., can be regarded as *derived quantities* and are expressed in terms of the fundamental dimensions. For example, the dimensions of force are given by ML/T^2, since force is equated to the product of mass and acceleration.

We could equally well regard force, length and time as the fundamental dimensions, using the symbols F, L, T. On this system, mass would be regarded as a derived quantity having dimensions FT^2/L. Some engineers prefer to work with the F, L, T system because it is claimed that force can be measured more directly than mass and is more easily understood as a fundamental concept. It does not really matter, however, which system is employed. The dimensions of some of the more common mechanical quantities are listed in the following table for purposes of reference both on the M, L, T and F, L, T systems:

Derived quantities	Dimensions		
Velocity	L/T		
Angular velocity	$1/T$		
Acceleration	L/T^2		
Mass	M	or	FT^2/L
Force	ML/T^2	or	F
Pressure	M/LT^2	or	F/L^2
Energy	ML^2/T^2	or	FL
Momentum	ML/T	or	FT
Density, ρ	M/L^3	or	FT^2/L^4
Specific weight	M/L^2T^2	or	F/L^3
Viscosity, μ	M/LT	or	FT/L^2
Kinematic viscosity, ν	L^2/T		
Surface tension, σ	M/T^2	or	F/L

The first important point to note is that, in any equation which expresses a certain relationship between a number of different physical quantities, each term in the equation must have the same dimensions. This is the principle of *dimensional homogeneity*. The main method of dimensional analysis, which is variously known as Thomson's theorem of numerics or as Buckingham's Π theorem, is a corollary of this principle and can be stated as follows:

If a physical relationship can be represented mathematically by expressing one variable Q_1 in terms of a number of other independent variables $Q_2, Q_3, Q_4, ...,$ etc., i.e. if

$$Q_1 = f(Q_2, Q_3, Q_4, ..., Q_n)$$

or $\qquad\qquad f(Q_1, Q_2, Q_3, ..., Q_n) = 0,$ \hfill (A)

the problem may be simplified by the following procedure:

(i) pick out p of the n Q's and regard them as primary Q's (where $p =$ number of fundamental dimensions required to describe the physical quantities involved), noting that each fundamental dimension must appear at least once among the dimensions of the primary Q's;

(ii) the remaining $(n - p)$ Q's are now expressed as $(n - p)$ dimensionless ratios, known as 'numerics' or 'Π quantities', Π_1, Π_2, Π_3, ... , Π_{n-p}, using the primary Q's for the purpose of forming the ratios, i.e. each Π quantity is formed from products or ratios of powers of Q's;

(iii) statement (A) may now be replaced by

$$\phi(\Pi_1, \Pi_2, \Pi_3, ..., \Pi_{n-p}) = 0.$$ \hfill (B)

The meaning of this theorem will be illustrated by a number of examples in the following sections.

5.2. Force or drag on a submerged body

If we confine ourselves to flow systems or models which have *geometrical similarity* we can say that the force exerted by a fluid stream flowing past a submerged body will depend on the physical properties of the fluid (e.g. density and viscosity), on the velocity of the stream, and on the linear scale or size of the body, i.e. we assume that force $F = $ a function of (ρ, μ, L, U) or

$$f(F, \rho, \mu, L, U) = 0.$$ \hfill (A)

In this case there are five Q's having dimensions

F	ρ	μ	L	U
ML/T^2	M/L^3	M/LT	L	L/T

i.e. $p = 3$ because we are concerned with purely mechanical phenomena here.

Various choices are possible for the primary Q's and they can be considered in turn:

(a) **First choice.** Take ρ, L, U as the primary Q's. We then require dimensionless ratios for F and μ. For Π_1 try $\dfrac{F}{\rho^{a_1} U^{b_1} L^{c_1}}$; checking the dimensions gives $a_1 = 1$, $b_1 = 2$, $c_1 = 2$, i.e. $\Pi_1 = \dfrac{F}{\rho U^2 L^2}$, which is known as a *force coefficient*, and by a similar argument $\Pi_2 = \dfrac{\mu}{\rho U L}$, which is the reciprocal of the *Reynolds number*.

The method of dimensional analysis therefore reduces the original statement (A) to the much simpler statement

$$\phi\left(\frac{F}{\rho U^2 L^2}, \frac{\mu}{\rho U L}\right) = 0, \tag{B}$$

i.e. we have reduced a problem involving five variables to one involving only two variables. Statement (B) can equally well be expressed as

$$\frac{F}{\rho U^2 L^2} = f_1\left(\frac{\rho U L}{\mu}\right), \tag{5.2.1}$$

i.e. the *force coefficient* is a function only of the *Reynolds number*.

(b) **Second choice.** Take μ, L, U as the primary Q's. We then require dimensionless ratios for F and ρ. Using the same method as before

$$\Pi_1 = \frac{F}{\mu U L} \quad \text{and} \quad \Pi_2 = \frac{\rho U L}{\mu}$$

so that

$$\frac{F}{\mu U L} = f_2\left(\frac{\rho U L}{\mu}\right). \tag{5.2.2}$$

$F/\mu U L$ is an alternative form of force coefficient so that once again force coefficient is a function of the Reynolds number.

(c) **Third choice.** Take ρ, μ, L as the primary Q's. We then require dimensionless ratios for F and U and we find that

$$\Pi_1 = \frac{F\rho}{\mu^2} \quad \text{and} \quad \Pi_2 = \frac{\rho U L}{\mu}$$

so that

$$\frac{F\rho}{\mu^2} = f_3\left(\frac{\rho U L}{\mu}\right). \tag{5.2.3}$$

$F\rho/\mu^2$ is yet another form for the force coefficient.

(d) Fourth choice. Taking ρ, μ, U as the primary Q's it is easily verified that the result is the same as in choice (c) above.

The physical significance of these different ratios will become clear in due course. It will be sufficient to note at this point that it is customary to select the ratio $\dfrac{F}{\rho U^2 L^2}$ or $\dfrac{F}{\frac{1}{2}\rho U^2 L^2}$ for general use as a force coefficient. Equation (5.2.1) is therefore the important result for this problem.

The validity of the method can best be proved by the following alternative approach. Considering the same problem of the force on a submerged body, we assume first of all that $F = f(\rho, \mu, L, U)$. Let us suppose that this functional relationship can be expressed mathematically as a series of i terms, each formed from powers of ρ, μ, L, U, i.e. that

$$F = \sum_i \alpha_i (\rho^{a_i} \mu^{b_i} L^{c_i} U^{d_i}),$$

α being a dimensionless coefficient. For each term we must have dimensionally

$$\frac{ML}{T^2} = \left(\frac{M}{L^3}\right)^{a_i} \left(\frac{M}{LT}\right)^{b_i} L^{c_i} \left(\frac{L}{T}\right)^{d_i},$$

i.e.
$$a_i + b_i = 1,$$
$$-3a_i - b_i + c_i + d_i = 1,$$
$$-b_i - d_i = -2;$$

we thus have three equations with four unknowns.

(a) First choice. Express a_i, c_i, d_i in terms of b_i:

$$a_i = 1 - b_i,$$
$$d_i = 2 - b_i,$$
$$c_i = 1 + 3a_i + b_i - d_i = 2 - b_i.$$
$$\therefore \quad F = \sum_i \alpha_i (\rho^{1-b_i} \mu^{b_i} L^{2-b_i} U^{2-b_i})$$

or
$$\frac{F}{\rho U^2 L^2} = \sum_i \alpha_i \left(\frac{\rho U L}{\mu}\right)^{-b_i} = f_1 \left(\frac{\rho U L}{\mu}\right), \qquad (5.2.1)$$

i.e. force coefficient is a function of the Reynolds number. Compare this procedure with the previous one.

(b) Second choice. Express b_i, c_i, d_i in terms of a_i:

$$b_i = 1 - a_i,$$
$$d_i = 2 - b_i = 1 + a_i$$
$$c_i = 1 + 3a_i + b_i - d_i = 1 + a_i.$$
$$\therefore \quad F = \sum_i \alpha_i (\rho^{a_i} \mu^{1-a_i} L^{1+a_i} U^{1+a_i})$$

or
$$\frac{F}{\mu U L} = \sum_i \alpha_i \left(\frac{\rho U L}{\mu}\right)^{a_i} = f_2 \left(\frac{\rho U L}{\mu}\right). \tag{5.2.2}$$

(c) Third choice. Express a_i, b_i, c_i in terms of d_i:

$$b_i = 2 - d_i,$$

$$a_i = 1 - b_i = d_i - 1,$$

$$c_i = 1 + 3a_i + b_i - d_i = d_i.$$

$$\therefore \quad F = \sum_i \alpha_i (\rho^{d_i - 1} \mu^{2 - d_i} L^{d_i} U^{d_i})$$

or
$$\frac{F\rho}{\mu^2} = \sum_i \alpha_i \left(\frac{\rho U L}{\mu}\right)^{d_i} = f_3 \left(\frac{\rho U L}{\mu}\right). \tag{5.2.3}$$

5.3. Physical significance of the Reynolds number and the force coefficient

The quantity ρU^2 is the flow of momentum through a stream tube of unit cross-sectional area. It can therefore be taken as a representative *inertia force per unit area*. If, for instance, the x-component of momentum was completely destroyed by the impingement of the stream against a solid wall or fixed body, the force exerted against the wall per unit area would be ρU^2.

The quantity U/L can be taken as giving an indication of the magnitude of the velocity gradients occurring in the flow. For instance, in flow through a pipe the velocity must range from zero at the wall of the pipe to a maximum value U_1 at the centre, and in this case we would take either the radius or the diameter of the pipe as the representative linear dimension L. Referring to §1.4, we can therefore say that the quantity $\mu U/L$ can be taken as a representative *shear stress* due to the action of viscosity.

The ratio $\dfrac{\rho U^2}{\mu U/L}$ or $\dfrac{\rho U L}{\mu}$, which is the *Reynolds number*, is therefore a measure of the relative magnitude of the inertial to the viscous forces occurring in the flow. The higher the Reynolds number the greater will be the relative contribution of inertia effects. The smaller the Reynolds number the greater will be the relative magnitude of the viscous stresses. We shall use the symbol Re for the Reynolds number.

The condition of *dynamical similarity* between two cases of incompressible flow with geometrically similar boundaries is satisfied if the Reynolds number has the same value in each case. A rigorous proof of this statement will be given in Chapter 12, §12.4.

In the example which was discussed in §5.2 the total force exerted on the submerged body was denoted by F. Taking L^2 as a representative area, it is evident that the force coefficient $F/\rho U^2 L^2$ is the ratio of the actual total force F to a representative inertia force $\rho U^2 L^2$.

The alternative form of force coefficient $F/\mu U L$ is the ratio of the actual total force F to a representative viscous force $\mu U L$ (viscous stress $\mu U/L$ multiplied by area L^2). The third form of force coefficient $F\rho/\mu^2$ does not have quite such an obvious physical meaning and is not normally used in practice. It is in fact related to the other two force coefficients in the following way:

$$\frac{F\rho}{\mu^2} = \left(\frac{F}{\mu U L}\right)^2 \bigg/ \frac{F}{\rho U^2 L^2}.$$

Note also that
$$\frac{F}{\rho U^2 L^2} = \frac{F}{\mu U L} \times \frac{1}{Re}.$$

5.4. Pressure drop for flow through a pipe

As another example of the method of dimensional analysis consider the problem of the flow of an incompressible gas or liquid through a pipe. Let us suppose that we wish to correlate the results of a large number of different experimental measurements of pressure drop. We start by assuming that

pressure drop $\Delta p =$ function of (ρ, μ, u_m, D, L),

or
$$f(\Delta p, \rho, \mu, u_m, D, L) = 0,$$

where u_m is the mean velocity in the pipe, D is the diameter, and L is the length of the pipe. We have six variables in this case and three fundamental dimensions. We should therefore be able to reduce the problem to a connexion between three dimensionless ratios.

Select ρ, u_m and D as the primary quantities. We then require dimensionless Π ratios for Δp, μ and L.

For Π_1 try
$$\frac{\Delta p}{\rho^{a_1} u_m^{b_1} D^{c_1}}.$$

Dimensionally
$$\frac{M}{LT^2} = \left(\frac{M}{L^3}\right)^{a_1} \left(\frac{L}{T}\right)^{b_1} L^{c_1},$$

hence,
$$a_1 = 1, \quad b_1 = 2, \quad c_1 = 0,$$

i.e.
$$\Pi_1 = \frac{\Delta p}{\rho u_m^2}, \text{ which is a pressure-drop coefficient.}$$

For Π_2 try
$$\frac{\mu}{\rho^{a_2} u_m^{b_2} D^{c_2}}.$$

Dimensionally $$\frac{M}{LT} = \left(\frac{M}{L^3}\right)^{a_2}\left(\frac{L}{T}\right)^{b_2} L^{c_2},$$

hence, $$a_2 = 1, \quad b_2 = 1, \quad c_2 = 1,$$

i.e. $\Pi_2 = \dfrac{\mu}{\rho u_m D}$, which is the reciprocal of the Reynolds number specified with the diameter of the pipe as the characteristic linear dimension.

Obviously $$\Pi_3 = L/D.$$

The conclusion therefore is that

$$\frac{\Delta p}{\rho u_m^2} = f\left(\frac{\rho u_m D}{\mu}, \frac{L}{D}\right). \tag{5.4.1}$$

For a long pipe, neglecting entry effects, it is found experimentally that Δp is directly proportional to the length L. We can therefore rewrite the last equation

$$\frac{\Delta p}{\rho u_m^2} = \frac{L}{D} f(Re) = 2c_f \frac{L}{D}, \tag{5.4.2}$$

where $2c_f = f(Re)$. The reason for this last substitution will become clear in Chapter 6. The quantity c_f, which is a function of the Reynolds number, is the *skin friction coefficient*.

If the Reynolds number $\rho u_m D/\mu$ has a numerical value less than about 2000, the flow in the pipe is generally found to be laminar. If, on the other hand, the Reynolds number is greater than 2000, the flow will normally be turbulent. This critical value of the Reynolds number was discovered experimentally by Osborne Reynolds in 1883, and since then it has been verified on innumerable occasions by other experimenters. There is as yet, however, no completely satisfactory theoretical explanation for the transition from laminar to turbulent flow. Further discussion of flow in pipes is reserved for Chapter 6.

5.5. Power required to drive a fan

As a further example of the method consider the case of a series of geometrically similar fans. Let N be the rotational speed, D the the impeller diameter and Q the volume flow. We will assume that the power P required to drive the fan will be a function of (ρ, μ, N, D, Q), or

$$f(P, \rho, \mu, N, D, Q) = 0.$$

We again have six variable and three fundamental dimensions, so that we should end up with three dimensionless ratios.

Select ρ, N, D as the primary quantities.

For Π_1 try $\dfrac{P}{\rho^{a_1} N^{b_1} D^{c_1}}$; checking the dimensions gives $a_1 = 1$, $b_1 = 3$, $c_1 = 5$, i.e. $\Pi_1 = \dfrac{P}{\rho N^3 D^5}$, which can be called a power coefficient.

For Π_2 try $\dfrac{\mu}{\rho^{a_2} N^{b_2} D^{c_2}}$, and we find $a_2 = 1$, $b_2 = 1$, $c_2 = 2$, i.e. $\Pi_2 = \dfrac{\mu}{\rho N D^2}$, which is the reciprocal of a Reynolds number.

For Π_3 try $\dfrac{Q}{\rho^{a_3} N^{b_3} D^{c_3}}$, and we find $a_3 = 0$, $b_3 = 1$, $c_3 = 3$, i.e. $\Pi_3 = \dfrac{Q}{N D^3}$, which can be called a flow coefficient.

The conclusion therefore is that

$$\frac{P}{\rho N^3 D^5} = f\left(\frac{Q}{N D^3}, \frac{\rho N D^2}{\mu}\right). \tag{5.5.1}$$

Note that we are still limiting ourselves to incompressible flow in this example.

Since the flow will be extremely turbulent owing to the passage of the air or gas through the fan blades viscous effects will usually be negligible compared with inertia effects. We can therefore ignore the variation with Reynolds number in this problem and say that for most practical purposes

$$\frac{P}{\rho N^3 D^5} = f\left(\frac{Q}{N D^3}\right). \tag{5.5.2}$$

5.6. Flow with a free surface

The classic case of the application of dimensional analysis to flow with a free surface is the problem of the resistance of a ship or floating body. We can no longer ignore gravity, and we must therefore bring the gravitational constant g into the picture as one of the physical quantities involved. We must now assume that the force F on the floating body can be expressed by the functional relationship

$$F = f(\rho, \mu, g, L, U),$$

where U is the velocity of flow past the body and L is the length. It may seem strange to regard the constant g as a variable, but even though it does not vary numerically in most engineering problems it is physically a relevant variable and it must be introduced as such if we are to find the correct dimensionless ratios.

Take ρ, L, U as the primary quantities and, applying the same technique as before, we find that

for F $\quad \Pi_1 = \dfrac{F}{\rho U^2 L^2}$, the *force coefficient*;

for μ $\quad \Pi_2 = \dfrac{\mu}{\rho U L}$, reciprocal of the *Reynolds number*;

for g $\quad \Pi_3 = \dfrac{gL}{U^2}$.

As in the case of the Reynolds number, it is customary to take the reciprocal of the last ratio Π_3. This is known as the *Froude number*, i.e. Froude number $Fr = U^2/gL$. The conclusion therefore is that

$$\frac{F}{\rho U^2 L^2} = f\left(\frac{\rho U L}{\mu}, \frac{U^2}{gL}\right). \tag{5.6.1}$$

It can be said that the variation of the force coefficient with Reynolds number represents the effect of skin friction, while the variation with Froude number represents the contribution of wave-making resistance.

If experiments are to be carried out on a model of a ship's hull, i.e. on a reduced linear scale, in order to get the correct value of the Froude number it will be necessary to employ a velocity for the model reduced in proportion to the square root of the ratio of the linear scale. But if this is done we will have the wrong value for the Reynolds number, unless of course we use a different liquid and thus alter the values of ρ and μ. If we confine ourselves to experiments with water we cannot satisfy the complete requirements of dynamical similarity, i.e. we cannot adjust matters so that Reynolds number *and* the Froude number have the same values for both the model and the full-scale ship. It is usual for experiments to be carried out with the correct value for the Froude number. The skin friction drag is estimated independently and is subtracted from the measured total drag of the model. We can thus obtain a coefficient for the wave-making drag alone. This is then used to calculate the wave-making resistance of the full-scale ship to which must be added the estimated skin friction drag.

The physical significance of the Froude number will be appreciated by writing it

$$Fr = \frac{U^2}{gL} = \frac{\rho U^2 L^2}{\rho g L^3}.$$

$\rho U^2 L^2$ is a representative inertia force and $\rho g L^3$ is a representative gravitational force. The Froude number therefore provides a measure of the ratio of inertia force to gravitational force in the flow problem considered.

5.7. Flow with surface tension

Consider next the problem of the flow of a liquid from an open-ended tube or nozzle leading to the formation of a spray of liquid drops. Suppose that we are interested in the average size of the drops formed in the spray. We will assume that

$$d_m = f(\rho, \mu, \sigma, g, D, U),$$

where d_m is the average diameter of the drops, U is the velocity in the tube, D is the diameter of the tube, σ is the surface tension of the liquid.

Let us take ρ, D, U as the primary quantities and we will then find that

$$\text{for } d_m \quad \Pi_1 = \frac{d_m}{D},$$

$$\text{for } \mu \quad \Pi_2 = \frac{\mu}{\rho U D}, \quad \text{reciprocal of the Reynolds number,}$$

$$\text{for } \sigma \quad \Pi_3 = \frac{\sigma}{\rho U^2 D}, \quad \text{reciprocal of the } \textit{Weber number We},$$

$$\text{for } g \quad \Pi_4 = \frac{gD}{U^2}, \quad \text{reciprocal of the Froude number.}$$

The conclusion therefore is that

$$\frac{d_m}{D} = f\left(\frac{\rho U D}{\mu}, \frac{\rho U^2 D}{\sigma}, \frac{U^2}{gD}\right), \tag{5.7.1}$$

or

$$d_m/D = f(Re, We, Fr).$$

CHAPTER 6

FLOW IN PIPES AND CHANNELS

6.1. Friction coefficients and the Reynolds number

Consider the steady flow of an incompressible gas or liquid in a long pipe of internal diameter D. Let the mean velocity be u_m. The *skin friction coefficient* c_f is defined by

$$\tau_0 = c_f \tfrac{1}{2}\rho u_m^2, \tag{6.1.1}$$

where τ_0 is the shear stress at the pipe wall.

Referring to fig. 33, it is evident that for a long pipe, neglecting entry effects, the pressure drop Δp over a length L is given by

$$\Delta p \frac{\pi D^2}{4} = \tau_0 \pi D L,$$

i.e.
$$\Delta p = 4 \frac{L}{D} \tau_0. \tag{6.1.2}$$

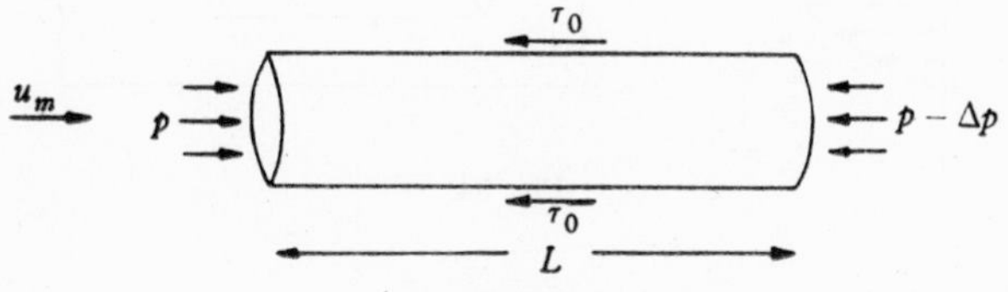

Fig. 33

Combining (6.1.1) and (6.1.2), the pressure drop may be expressed by

$$\Delta p = 2 c_f \frac{L}{D} \rho u_m^2. \tag{6.1.3}$$

This is sometimes called the Fanning equation, although it is in fact little more than a definition of the skin friction coefficient c_f and hardly merits the title of an equation. We have already arrived at this result in §5.4, using the methods of dimensional analysis.

If we prefer to deal in terms of head rather than pressure, we can say that the loss of head is given by

$$\frac{\Delta p}{\rho g} = 2 c_f \frac{L}{D} \frac{u_m^2}{g}. \tag{6.1.4}$$

It was shown in §5.4 that the skin friction coefficient c_f is a function of the Reynolds number. This is true for flow in *smooth pipes*.

6   65

In general, however, c_f will also depend on the roughness of the pipe wall and we must therefore say that

$$c_f = f(Re, \text{roughness}), \qquad (6.1.5)$$

where $Re = \rho u_m D / \mu$, and roughness has yet to be defined.

For non-circular pipes and channels it is convenient to define the 'hydraulic radius' m as the ratio of the cross-sectional flow area to the wetted perimeter. We can then continue to use equations (6.1.3), (6.1.4) and (6.1.5) if we put $4m$ in place of D.

6.2. Laminar flow in pipes

If the Reynolds number is less than about 2000 the flow in a smooth pipe will generally be laminar. For fully developed viscous

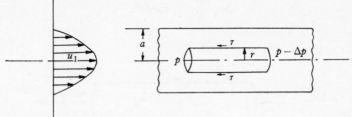

Fig. 34

flow in a long pipe the theoretical velocity distribution is parabolic and this is verified by experiment. The theoretical velocity profile can be derived as follows:

Referring to fig. 34, consider the motion of a cylindrical element of fluid of radius r and length L. The total viscous force acting on the cylindrical element is

$$2\pi r L \tau = -2\pi r L \mu \frac{du}{dr};$$

therefore, for steady motion of the element,

$$\pi r^2 \Delta p = -2\pi r L \mu \frac{du}{dr}$$

and

$$-\int_{u_1}^{0} du = \frac{\Delta p}{2\mu L} \int_{0}^{a} r \, dr,$$

since $u = u_1$ (the maximum velocity) at $r = 0$, and $u = 0$ at $r = a$.

Therefore the maximum velocity at the centre of the pipe is given by

$$u_1 = \frac{\Delta p}{4\mu L}\, a^2, \tag{6.2.1}$$

and the velocity distribution is given by

$$u = \frac{\Delta p}{4\mu L}\, (a^2 - r^2). \tag{6.2.2}$$

The total volume flow Q is given by

$$Q = \int_0^a 2\pi\, ru\, \mathrm{d}r = \frac{\Delta p}{8\mu L}\, \pi a^4, \tag{6.2.3}$$

and the mean velocity u_m is given by

$$u_m = \frac{Q}{\pi a^2} = \frac{\Delta p}{8\mu L}\, a^2 = \frac{u_1}{2}. \tag{6.2.4}$$

The last result may be rewritten

$$\Delta p = \frac{8\mu u_m L}{a^2} = \frac{32\mu u_m L}{D^2}; \tag{6.2.5}$$

this is the *Hagen-Poiseuille law* for the pressure drop in a long pipe with laminar viscous flow.

From (6.1.3) and (6.2.5) the skin friction coefficient c_f for laminar flow in a long pipe can be expressed as

$$c_f = \frac{16\mu}{\rho u_m D} = \frac{16}{Re}, \tag{6.2.6}$$

which is a special case of the general statement of (6.1.5).

None of the preceding results can be applied to the entry region of a pipe. The form of the velocity profiles in the entry region is shown diagrammatically in fig. 35. At the start the velocity will be uniform across the section of the pipe with a value which must be equal, by the principle of continuity, to the mean velocity u_m for the fully developed parabolic profile. Owing to the action of viscosity a *boundary layer* will form on the wall of the pipe and will gradually spread inwards towards the centre. The central core of fluid, which is otherwise unaffected by the action of viscosity, will gradually be accelerated from the initial velocity u_m to the final maximum value u_1 when the boundary layer converges to the centre-line of the pipe and thus embraces the entire flow. Theoretically the approach to the fully developed parabolic velocity profile is asymptotic. An approximate estimate of the magnitude of the entry length ([1], Chapter VII, §139), however, gives the following result:

$$L'/D = 0{\cdot}029Re, \qquad (6.2.7)$$

and this is found to be in good agreement with experimental results.

For the final parabolic profile, from (6.2.1) and (6.2.2)

$$\frac{u}{u_1} = 1 - \frac{r^2}{a^2},$$

and the mean kinetic energy is given by

$$\frac{1}{\pi a^2 u_m} \int_0^a 2\pi r \rho u \left(\frac{u^2}{2}\right) \mathrm{d}r = \frac{8\rho u_m^2}{a^2} \int_0^a \left(1 - \frac{r^2}{a^2}\right)^3 r \, \mathrm{d}r = \rho u_m^2,$$

i.e. the final mean kinetic energy is equal to twice the kinetic energy at the entry to the pipe. The total pressure drop associated with the acceleration of the fluid is therefore ρu_m^2 and not $\frac{1}{2}\rho u_m^2$.

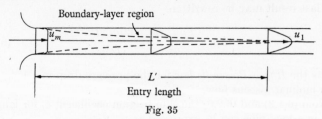

Fig. 35

We have seen from (6.2.5) that for fully developed laminar flow the pressure drop is proportional to the mean velocity u_m. In the entry region, however, the pressure drop varies more rapidly with u_m owing to the changes which occur in the velocity profile in this region.

6.3. Turbulent flow in a smooth pipe

Laminar flow will normally break down if the Reynolds number $Re > 2000$ approximately. For turbulent flow in smooth pipes, for values of Re up to 10^5, the observed pressure drop is such as to imply a value of the skin-friction coefficient c_f given by

$$c_f = 0{\cdot}079 Re^{-\frac{1}{4}}, \qquad (6.3.1)$$

and this is another special case of the general statement of (6.1.5). The formula of (6.3.1) is known as the Blasius law[2].

The shear stress τ_0 at the wall and the pressure drop under these conditions are given by

$$\tau_0 = c_f \tfrac{1}{2}\rho u_m^2 = 0.0395\rho v^{\frac{1}{4}} D^{-\frac{1}{4}} u_m^{\frac{7}{4}}, \qquad (6.3.2)$$

and $\qquad \Delta p = 2c_f \dfrac{L}{D} \rho u_m^2 = 0.158 L\rho v^{\frac{1}{4}} D^{-\frac{5}{4}} u_m^{\frac{7}{4}}, \qquad (6.3.3)$

i.e. the pressure drop varies as $u_m^{\frac{7}{4}}$.

For higher values of the Reynolds number, beyond 10^5, a more recent formula for the skin friction coefficient c_f is

$$\frac{1}{\sqrt{c_f}} = 4.0 \log_{10}(Re\sqrt{c_f}) - 0.4, \qquad (6.3.4)$$

and this is known as the Kármán-Nikuradse formula. It has some slight theoretical foundation which will be discussed in Chapter 14.

Equation (6.3.4) is inconvenient to use because one cannot solve it directly for c_f. The two formulae (6.3.1) and (6.3.4) are therefore plotted graphically in fig. 36. The following table of values of c_f calculated from (6.3.4) may also be useful:

Re	c_f
2×10^4	0.00650
5×10^4	0.00522
10^5	0.00449
2×10^5	0.00391
5×10^5	0.00329
10^6	0.00292
2×10^6	0.00260
5×10^6	0.00226
10^7	0.00204

Note carefully that these results only apply to smooth pipes.

The velocity distribution for fully developed turbulent flow in a pipe will be discussed in some detail in Chapter 14. We can, however, draw some conclusions at this stage from the simple empirical Blasius expression for the shear stress τ_0 at the wall represented by equation (6.3.2).

We have already noted that in the entry region of a pipe a boundary layer is formed on the wall and that this gradually spreads inwards as the fluid moves along the pipe. Near the entry, with laminar flow, the velocity profile is of the form shown in fig. 37(a), consisting of a laminar boundary layer and a central core.

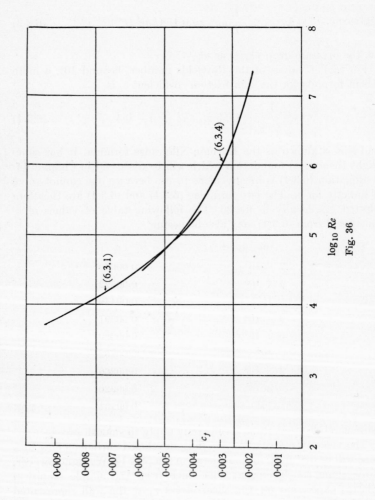

Fig. 36

At large values of the Reynolds number transition will occur from laminar to turbulent flow within the boundary layer at a point close to the entry of the pipe resulting in a velocity profile of the form shown in fig. 37 (*b*). This turbulent boundary layer will then spread inwards as the flow continues downstream. We will assume that for the turbulent boundary layer, when it is fully developed,

$$\frac{u}{u_1} = f\left(\frac{y}{a}\right) = \left(\frac{y}{a}\right)^n, \tag{6.3.5}$$

where y = distance from the wall = $(a - r)$.

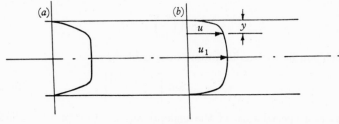

Fig. 37

The mean velocity u_m will have a constant ratio to the maximum velocity at the centre u_1 depending on the value of the exponent n, i.e.

$$u_m = \text{constant} \times u_1 = \text{constant} \times \left(\frac{a}{y}\right)^n u = \text{constant} \times \left(\frac{D}{y}\right)^n u,$$

but from (6.3.2) $\qquad \tau_0 = 0{\cdot}0395\,\rho v^{\frac{1}{4}} D^{-\frac{1}{4}} u_m^{\frac{7}{4}},$

and therefore $\qquad \tau_0 = \text{constant} \times \rho v^{\frac{1}{4}} D^{\frac{7}{4}n - \frac{1}{4}} y^{-\frac{7}{4}n} u^{\frac{7}{4}}$

or $\qquad u = \text{constant} \times \left(\frac{\tau_0}{\rho}\right)^{\frac{4}{7}} v^{-\frac{1}{7}} D^{\frac{1}{7} - n} y^n. \tag{6.3.6}$

Near the wall, where the flow is hardly distinguishable from the flow in a true boundary layer, however, we would not expect the velocity distribution to depend on the pipe diameter D. This statement can to some extent be justified by the fact that turbulent velocity profiles are observed to be relatively flat at the centre of the pipe as suggested in the diagram of fig. 37 (*b*), and that large changes of velocity are confined to a region fairly close to the wall. In this region, at any rate, if u is to be independent of D it is evident from (6.3.6) that we must have $n = \frac{1}{7}$, i.e. the velocity profile fairly close to the wall should be given by the one-seventh power law

$$\frac{u}{u_1} = \left(\frac{y}{a}\right)^{\frac{1}{7}}. \tag{6.3.7}$$

It is observed in practice that this does give a tolerably good approximation to the measured velocity profile, not only near the wall but also over the central portion of the flow, in the range of values of the Reynolds number up to about 10^5 for which the Blasius law (6.3.1) or (6.3.2) is valid. At higher values of the Reynolds number the observed velocity profiles correspond to successively lower values of the exponent n in the power law of (6.3.5).

In the range of Reynolds number up to 10^5, if we take $n = \frac{1}{7}$, the mean velocity for the profile of (6.3.7) will be related to the maximum velocity at the centre by $u_m = 0.817u_1$, and equation (6.3.6) becomes

$$u = 8.6 \left(\frac{\tau_0}{\rho}\right)^{\frac{4}{7}} \nu^{-\frac{1}{7}} y^{\frac{1}{7}},$$

which can be written in a more intelligible form as

$$\frac{u}{\sqrt{(\tau_0/\rho)}} = 8.6 \left[\frac{y \sqrt{(\tau_0/\rho)}}{\nu}\right]^{\frac{1}{7}}. \qquad (6.3.8)$$

This is a special case of the general relationship for the velocity distribution near a wall with turbulent flow which will be discussed in Chapter 14:

$$\frac{u}{\sqrt{(\tau_0/\rho)}} = f\left(\frac{y \sqrt{(\tau_0/\rho)}}{\nu}\right). \qquad (6.3.9)$$

The quantity $\sqrt{(\tau_0/\rho)}$ which appears in these expressions has the dimensions of velocity and is referred to as the *friction velocity*. Another special case of (6.3.9) is the so-called 'universal velocity profile' which is applicable to a wider range of Reynolds number,

$$\frac{u}{\sqrt{(\tau_0/\rho)}} = 5.5 + 5.75 \log_{10}\left(\frac{y \sqrt{(\tau_0/\rho)}}{\nu}\right). \qquad (6.3.10)$$

The Kármán-Nikuradse formula (6.3.4) for the skin friction coefficient bears essentially the same relationship to the velocity profile of (6.3.10) as does the Blasius formula (6.3.1) to the profile of (6.3.8).

It should be mentioned at this stage that although we speak of these turbulent velocity profiles applying in a region fairly close to the wall of the pipe, they cannot continue to be valid right up to the wall itself. For instance, the profile of (6.3.7) or (6.3.8) would imply an infinite velocity gradient at the wall where $y = 0$. This cannot be true because the mechanism by which the turbulent shearing stress is finally transmitted to the wall of the pipe must be that of viscosity. We must, in other words, have some form of *laminar sub-layer* adjacent to the wall with a finite velocity gradient corresponding to a finite shearing stress. There is some degree of uncertainty regarding the exact nature and thickness of this laminar sub-layer, but most of the evidence points towards an order of magnitude for the thickness of the sub-layer δ_L given by

$$\frac{\delta_L \sqrt{(\tau_0/\rho)}}{\nu} \approx 10. \tag{6.3.11}$$

Further discussion of turbulent velocity profiles and of the underlying theoretical considerations will be deferred until Chapters 13 and 14. In the meantime the simple power-law profile of (6.3.7) may be taken as giving a reasonable approximation for the velocity distribution outside the laminar sub-layer provided the Reynolds number is not greater than about 10^5.

Note that the mean kinetic energy of the flow for the profile of (6.3.7) is given by

$$\frac{1}{\pi a^2 u_m} \int_0^a 2\pi r\rho \, \frac{u^3}{2} \, \mathrm{d}r = 1 \cdot 058 \, \frac{\rho u_m^2}{2},$$

which is only slightly different from $\frac{1}{2}\rho u_m^2$.

6.4. Turbulent flow in a rough pipe

Most commercial pipes are not in fact smooth, and the values for c_f calculated from (6.3.1) or (6.3.4) represent only the lower limit for the friction loss at the Reynolds number specified. Experiments have been carried out with artificially roughened pipes using sand grains, or other solid particles of selected size, glued to the wall of an otherwise smooth pipe. It is then possible to quote a definite *roughness size* ε corresponding to the diameter of the particles. It is found that the values of c_f calculated from the experimental measurements of the pressure drop can be correlated, for large values of the Reynolds number, by the formula

$$\frac{1}{\sqrt{c_f}} = 4 \cdot 0 \log_{10}\left(\frac{a}{\varepsilon}\right) + 3 \cdot 46, \tag{6.4.1}$$

which is the counterpart to (6.3.4) for smooth pipes.

Equation (6.4.1) represents the extreme case of rough pipes at large Reynolds numbers when the skin friction coefficient is a function only of the relative roughness and is independent of the Reynolds number. In these circumstances the pressure drop is proportional to u_m^2.

The surface of a commercial pipe, however, differs from that of an artificially roughened pipe and cannot be described in terms of a single roughness size. We need an average equivalent particle size together with a shape factor and a size distribution to describe the surface properly. A satisfactory method of correlation has not yet been found which will enable reliable predictions to be made for widely different types of commercial pipe. In general, the skin friction coefficient will depend both on the relative roughness and on the Reynolds number, so that most practical cases lie somewhere between the two extremes of the smooth pipe law of (6.3.1) or (6.3.4) on the one hand and the rough pipe law of (6.4.1) on the other. The variation of pressure drop with velocity will therefore follow a relationship which is intermediate between Δp being proportional to $u_m^{1.75}$ for the case of smooth pipes and Δp being proportional to

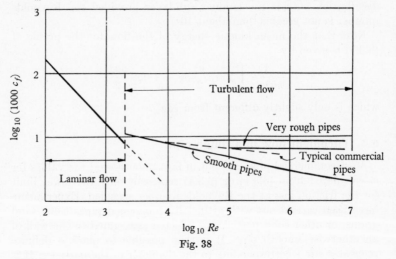

Fig. 38

u_m^2 for very rough pipes. This is illustrated diagrammatically in fig. 38, but the reader should consult [3], [4] and [5] for advice on practical formulae.

6.5. Energy and hydraulic gradients

The loss of head due to friction in a long pipe-line can be calculated from (6.1.4) provided we know the correct value for the skin friction coefficient c_f. In addition to the frictional loss there will generally be other losses of an inertial nature due to sudden changes of section, flow through valves or fittings, and flow round bends. The loss of head due to break-away and eddying effects in a sudden expansion, for instance, has already been estimated in §4.1. If we now consider a long pipe-line discharging from a reservoir, such as that illustrated in fig. 39, we could estimate the total loss of head h_f due to friction and eddying effects incurred up to any section of the pipe, and we

could set off this height below the level of the free surface in the
reservoir. The level thus arrived at, measured above the datum level,
represents the total head available at that section of the pipe-line,
i.e. the sum of the static head represented by the level of the pipe
itself together with the pressure head and velocity head of the fluid
in the pipe. A line drawn through these levels is known as the
energy gradient.

If we now picture the pipe-line fitted with a number of open-ended
stand pipes the level of the fluid in each stand pipe would represent
the sum of the static head and pressure head at that section of the
pipe-line. A line drawn through the levels in the stand pipes is
known as the *hydraulic gradient*. The vertical distance between the

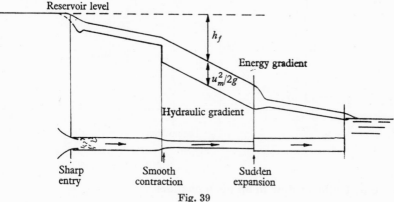

Fig. 39

energy and hydraulic gradient lines then represents the velocity
head in the pipe-line at each point. Fig. 39 shows the two lines
drawn for a horizontal pipe-line, with various changes of section,
discharging from a reservoir. Fig. 40 shows the same diagrammatic
method applied to a pipe-line with changes of level. Note that the
energy gradient line always falls but that the hydraulic gradient
line can go either up or down. If the hydraulic gradient line falls
below the level of the pipe-line at any point in its course it means
that the pressure inside the pipe will be below atmospheric at that
point. This state of affairs is undesirable because of the danger
of entrainment of air and of cavitation.

Fig. 40 is intended to represent a portion of a long pipe-line
between two pumping stations at A and G. The diagram shows a
gradient profile for the pipe-line. It may be assumed in this particular case that the vertical scale is exaggerated and that the frictional loss of head h_f may be taken as being proportional to the

horizontal distance without serious error. It is also assumed that the cross-sectional area of the pipe does not change and that the mean velocity u_m therefore remains constant. With these provisos the hydraulic gradient may be drawn as a straight line. It will be apparent from the figure that the static pressure inside the pipe would fall below atmospheric pressure around point E, so that the arrangement shown would not in fact be satisfactory.

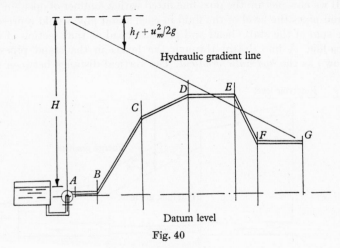

Fig. 40

6.6. Uniform flow in an open channel

For steady flow in an open channel the hydraulic gradient line will coincide by definition with the free surface of the flowing liquid. If, in addition, the flow is *uniform*, i.e. the velocity does not change with distance along the channel, the free surface will be parallel to the channel floor and the loss of head per unit length will therefore be equal to the slope of the channel floor.

Considering first of all the case of two-dimensional flow near the centre of a very wide channel, and referring to fig. 41, we can say that for an element of fluid of length dx and depth $(\lambda - y)$

$$\tau \, dx = (\lambda - y)\rho g \sin \alpha \, dx,$$

where τ is the shearing stress at distance y from the floor; hence

$$\tau = -(\lambda - y)\rho g \frac{dh_0}{dx}. \qquad (6.6.1)$$

At $y = 0$ $\qquad \qquad \tau = \tau_0 = c_f \tfrac{1}{2}\rho u_m^2,$

and therefore
$$- \lambda \rho g \frac{\mathrm{d}h_0}{\mathrm{d}x} = c_f \tfrac{1}{2} \rho u_m^2,$$

i.e.
$$- \frac{\mathrm{d}h_0}{\mathrm{d}x} = \frac{c_f u_m^2}{2g\lambda}. \tag{6.6.2}$$

Note that the velocity will be a maximum at the surface where $\tau = 0$ by equation (6.6.1). This is strictly true, however, only for a

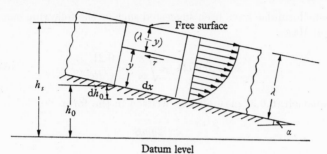

Fig. 41

channel of infinite width. For a channel of finite cross-section the maximum velocity will occur slightly below the surface owing to secondary flow effects.

For a channel of finite width, such as the cross-section shown in fig. 42, it is necessary to define the *hydraulic mean depth m* as the ratio of the cross-sectional area of flow to the wetted perimeter. Note

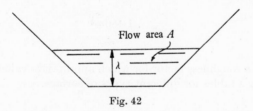

Fig. 42

that the free surface is not included in the wetted perimeter. We can then say approximately, for steady uniform flow, that

$$\tau_0 \frac{A}{m} \, \mathrm{d}x = \rho g \sin \alpha \, A \, \mathrm{d}x = - \rho g A \, \mathrm{d}h_0;$$

therefore
$$\tau_0 = - \rho g m \frac{\mathrm{d}h_0}{\mathrm{d}x} = c_f \tfrac{1}{2} \rho u_m^2,$$

and
$$- \frac{\mathrm{d}h_0}{\mathrm{d}x} = \frac{c_f u_m^2}{2gm}. \tag{6.6.3}$$

In all practical cases of open channels the flow is turbulent. Strictly speaking the friction coefficient c_f will be a function of the Reynolds number, the relative roughness, and the shape of the cross-section, i.e.

$$c_f = f\left(\frac{\rho u_m m}{\mu},\ \frac{\varepsilon}{m},\ \text{shape factor}\right). \tag{6.6.4}$$

Various formulae have been suggested such as the following one due to von Mises:

$$c_f = 0.0024 + \sqrt{\frac{\varepsilon}{m}} + \frac{0.21}{\sqrt{Re}}. \tag{6.6.5}$$

Equation (6.6.3) may also be written in the form

$$u_m^2 = \frac{2gmS}{c_f},$$

where $S = -\,dh_0/dx = $ slope of channel floor, or

$$u_m = C\sqrt{(mS)}, \tag{6.6.6}$$

where $C = \sqrt{(2g/c_f)}$. This is the *Chezy formula* which is widely used by civil engineers. Tables and formulae have been compiled giving values of the Chezy coefficient C. For instance, the Manning formula gives [6]

$$C = \frac{1.486 m^{\frac{1}{6}}}{n}, \tag{6.6.7}$$

where n is a roughness factor for which appropriate values are given in Manning's tables for different types of surface.

6.7. Loss of head with turbulent flow through pipe bends and fittings

In addition to the frictional loss in a pipe there may be other losses, as we have already noted in discussing the energy gradient, associated with flow through pipe bends and fittings. In many process plants these other losses outweigh the frictional loss. It is convenient to express these losses in terms of the velocity head $u_m^2/2g$. The following table will give some idea of their magnitude, but reference should be made to the various engineering handbooks

and to data supplied by the manufacturers of valves and fittings
for actual design calculations on pipework:

Typical loss of head

Sharp right-angle bend	1·0 to 1·2 $u_m^2/2g$
Commercial elbows	0·6 to 0·8 ⎫
Rounded bends	0·2 to 0·3 ⎬ $u_m^2/2g$ depending on radius
Gate valve fully open	0·2 $u_m^2/2g$
Globe valve	up to 10 $u_m^2/2g$

In each case u_m is the *mean velocity in the connecting pipe.*

The nature of the flow through a pipe bend is complicated by the
fact that a *secondary flow* is superimposed on the mean velocity.

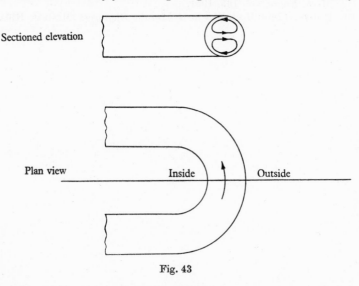

Sectioned elevation

Plan view   Inside   Outside

Fig. 43

Referring to fig. 43, it will be apparent that a pressure gradient
must be set up across the pipe to balance the centrifugal forces. The
fluid near the top and bottom of the section shown in the diagram
moves more slowly than the fluid at the centre, and a smaller
pressure gradient is sufficient to balance the centrifugal force. The
result of this is that there must be an outward movement of the
fluid at the centre and an inward movement of the fluid at the top
and bottom superimposed on the ordinary flow along the axis of the
pipe. The region of maximum velocity is thus displaced outwards
from the centre of the pipe. The resulting distortion of the velocity
profile may survive for a considerable distance after the bend has

been passed. It might almost be said that there are few instances in process plants where the length of run of a straight pipe is sufficient to even out these distortions of the velocity profile.

REFERENCES

(1) GOLDSTEIN (editor). *Modern Developments in Fluid Dynamics* (Oxford University Press).
(2) PRANDTL. *The Essentials of Fluid Dynamics* (Blackie).
(3) MOODY. Friction factors for pipe flow. *Trans. Amer. Soc. Mech. Engrs*, A.S.M.E. vol. 66, 1944.
(4) BLAIR. New formulae for water flow pipes. *Proc. Instn Mech. Engrs*, vol. 165, 1951.
(5) PROSSER *et al.* Friction losses in turbulent pipe flow. *Proc. Instn Mech. Engrs*, vol. 165, 1951.
(6) ROUSE. *Fluid Mechanics for Hydraulic Engineers* (McGraw-Hill).

CHAPTER 7

PUMPS AND COMPRESSORS

7.1. Classification of pumps

Pumps can be divided into two main categories, *positive displacement pumps*, whether of the reciprocating or rotary type, on the one hand, and centrifugal or axial-flow *impeller pumps* on the other.

The characteristic feature of a *positive displacement* pump is that it will endeavour to deliver a definite volume of liquid at every stroke or revolution regardless of the head against which it is required to work. There are two common types of reciprocating pump, the double-acting piston pump, and the single-acting plunger pump. There are a large number of different designs of rotary positive displacement pump, of which perhaps the gear pump is the best known. Generally speaking positive displacement pumps of one type or another are employed where it is required to deliver a relatively small volume of liquid against a relatively high head. The principles of fluid dynamics scarcely enter into their design, since the volume flow is determined by the geometry of the pump and is given by the swept volume per stroke less the leakage rate, and the head to be developed is fixed by the system or reservoir into which the pump is delivering and not by the pump itself. This is not to say that problems of mechanics and machine design do not arise. Clearly the pump must be strong enough mechanically to withstand the maximum pressures and stresses developed within it at any time. Some form of relief valve or by-pass valve will normally be fitted to prevent damage in the event, for instance, of a sudden blockage in the delivery system.

An *impeller pump*, on the other hand, whether of the centrifugal or axial-flow type, functions on an entirely different principle. The maximum head that can be developed is a definite characteristic of the particular design of pump, while the volume flow can vary from zero to a certain maximum value corresponding to zero delivery head. The essential components of a *centrifugal pump* are an impeller, usually having curved blades, mounted on a shaft and running inside a volute casing. The fluid enters axially into the eye of the impeller and is discharged tangentially from the volute. Some centrifugal pumps have a diffuser ring consisting of a number of fixed guide vanes located at the outer edge of the impeller inside the spiral casing. These are sometimes known as *diffuser pumps* in

81

contrast to *volute pumps*. Centrifugal pumps may be built either as single-stage or multi-stage machines. In the latter case a number of impellers are mounted on a single shaft and function in series. Generally speaking centrifugal pumps are employed where it is required to deliver a relatively large volume flow against a medium head. *Axial-flow* pumps are used for handling still larger volume flows and delivering against low heads. *Mixed-flow* pumps cover a range of duties intermediate between radial-flow centrifugal pumps at one end of the scale and axial-flow pumps at the other. The appropriate ranges of application of different types of pump, however, will be given quantitatively in §7.6.

One of the biggest problems in designing a pump, especially for chemical plant duties, is to provide an effective means of sealing the liquid inside the pump and preventing it from leaking out where the shaft emerges from the casing. This is a particularly serious problem in the case of liquids which are corrosive, inflammable, or toxic. Various types of shaft seal are employed ranging from the conventional stuffing box and gland to the so-called mechanical seal. In the conventional gland, if the pump has a suction pressure below atmospheric, a lantern ring is usually provided at the centre of the stuffing box acting as a spacer which divides the packing effectively into two parts. An external pressure feed of liquid is maintained on the stuffing box by means of a pipe connexion at the position occupied by the lantern ring. There will then be a small leakage of liquid past the packing, both inwards towards the pump impeller and outwards to atmosphere. The pressure feed either can be taken from the delivery side of the pump casing, or it may be an entirely separate supply using a gland-sealing fluid different from the main fluid which is being handled by the pump. In a mechanical seal rubbing contact is maintained, usually between a carbon ring and some specially hard metallic surface. Even with a rubbing gland, however, there may be some leakage. It is safe to say that in any conventional gland there must either be a slight outward leakage of the fluid which is being pumped, or a slight inward leakage of some gland sealing fluid.

Recently *totally enclosed pumps* have been developed for certain special applications. These usually incorporate a totally enclosed electric motor of the induction type with the rotor mounted directly on the pump shaft and with bearings which are designed to run in the fluid which is being handled by the pump. There is then no mechanical connexion between the pump impeller and the outside world so that the casing can be completely sealed.

In the following sections attention will be concentrated mainly on the centrifugal pump.

7.2. Flow through a centrifugal pump impeller

The flow of liquid through a centrifugal pump impeller is illus-trated diagrammatically in fig. 44 with the help of velocity dia-grams drawn for the inlet and outlet radii. We will suppose that the velocity triangles represent the *actual mean velocities* at these radii. For reasons which will shortly become apparent, a pump impeller normally has backward-swept curved blades as indicated in fig. 44. We cannot say, however, that the relative velocity of the fluid will be exactly parallel to the blades. In practice it is found that the angles β_1' and β_2' in the velocity triangles of fig. 44 will differ from the blade angles β_1 and β_2. This difference is particularly marked in the case of the outlet velocity triangle. More will be said on this point in a moment. Provided we have the actual measured velocities recorded on the diagram, however, we can proceed to write down the angular momentum equation. The notation is the same as that used for the turbine runner in §4.5, i.e. u denotes the tangential velocity of the impeller, v_r represents the relative velocity of the fluid, while the absolute velocity v is resolved into two components, a velocity of whirl w, and a velocity of flow f. Taking a control surface which encloses the impeller, and applying the angular momentum equation, we can say that the torque exerted by the impeller on the fluid is equal to the net rate of outflow of angular momentum from the control surface, i.e.

$$T = \rho Q(r_2 v_2 \cos \alpha_2 - r_1 v_1 \cos \alpha_1)$$

or
$$T = \rho Q(r_2 w_2 - r_1 w_1). \tag{7.2.1}$$

The power input will be given by

$$T\Omega = \rho Q(w_2 u_2 - w_1 u_1), \tag{7.2.2}$$

and the work input per unit mass flow will be

$$W_i = (w_2 u_2 - w_1 u_1). \tag{7.2.3}$$

The theoretical gain of total head in the impeller is

$$H_i = \frac{(w_2 u_2 - w_1 u_1)}{g}. \tag{7.2.4}$$

If there is no pre-rotation of the fluid entering the eye of the impeller, i.e. if the velocity of whirl $w_1 = 0$,

$$H_i = \frac{w_2 u_2}{g}, \tag{7.2.5}$$

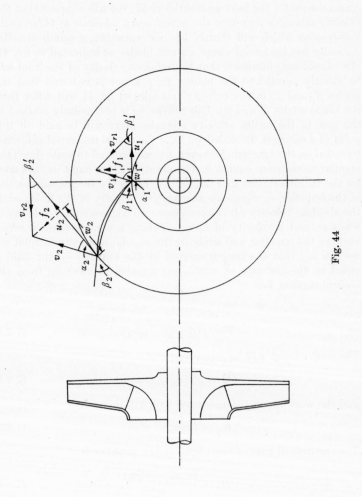

Fig. 44

and this can be written

$$H_i = \frac{u_2(u_2 - f_2 \cot \beta_2')}{g}.$$

In practice we do not normally know the numerical value of the angle β_2', although we do know the blade angle β_2. We could therefore define a *virtual head* H_v by the statement

$$H_v = \frac{u_2(u_2 - f_2 \cot \beta_2)}{g}, \qquad (7.2.6)$$

and say that the actual input head H_i is related to the virtual head H_v by $H_i = \eta_v H_v$, where η_v is the *vane efficiency*. The virtual head H_v is the head that would be gained by the fluid if the relative velocity at exit was exactly parallel to the blades of the impeller, i.e. if the angle β_2' was equal to β_2. H_v is also known variously as the

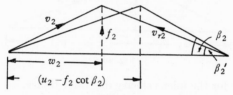

Fig. 45

ideal head or the Euler head. It is part of the art of pump design to be able to forecast the relationship between H_i and H_v. Another way of putting this matter is to say that, owing to secondary flow effects in the impeller, the actual whirl velocity w_2 will be less than the virtual or ideal whirl velocity $(u_2 - f_2 \cot \beta_2)$ and will be related by

$$w_2 = \eta_v(u_2 - f_2 \cot \beta_2), \qquad (7.2.7)$$

where the factor η_v will depend on the number of vanes but will usually have a value around 0·7. The geometrical relationship between the actual and virtual whirl velocities is shown in fig. 45. It will be clear from the diagram that the angle β_2' will always be less than the blade angle β_2.

7.3. The energy equation for a centrifugal pump

We can apply the *mechanical energy equation* to the case of flow through a centrifugal pump as follows:

Flow through the impeller:

$$\frac{p_2}{\rho} + \frac{v_2^2}{2} = \frac{p_1}{\rho} + \frac{v_1^2}{2} + W_i - L_i, \qquad (7.3.1)$$

where L_i is the mechanical energy loss in the impeller per unit mass flow due to friction and eddying.

Substituting for W_i from (7.2.3), this can be written

$$\frac{p_2 - p_1}{\rho} = \left(w_2 u_2 - \frac{v_2^2}{2}\right) - \left(w_1 u_1 - \frac{v_1^2}{2}\right) - L_i.$$

But, referring to fig. 46,

$$v_2^2 = w_2^2 + f_2^2 = w_2^2 + v_{r2}^2 - (u_2 - w_2)^2$$
$$= v_{r2}^2 - u_2^2 + 2w_2 u_2,$$

and therefore

$$w_2 u_2 - \frac{v_2^2}{2} = \tfrac{1}{2}(u_2^2 - v_{r2}^2),$$

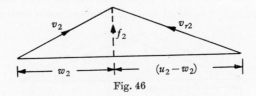

Fig. 46

and similarly for the inlet velocity diagram. Hence,

$$\frac{p_2 - p_1}{\rho} = \frac{(u_2^2 - u_1^2)}{2} - \frac{(v_{r2}^2 - v_{r1}^2)}{2} - L_i$$

or

$$\frac{p_2 - p_1}{\rho} = \frac{\Omega^2}{2}(r_2^2 - r_1^2) - \frac{(v_{r2}^2 - v_{r1}^2)}{2} - L_i. \tag{7.3.2}$$

If there is no flow, i.e. if the pump is running with the delivery valve closed,

$$\frac{p_2 - p_1}{\rho} = \frac{\Omega^2}{2}(r_2^2 - r_1^2), \tag{7.3.3}$$

which is the usual expression for a *forced vortex*. Compare (2.6.4).

Flow through the diffuser or volute casing:

$$\frac{p_3}{\rho} + \frac{v_3^2}{2} = \frac{p_2}{\rho} + \frac{v_2^2}{2} - L_d, \tag{7.3.4}$$

where suffix 3 refers to outlet from the diffuser or volute, and L_d is the mechanical energy loss in the diffuser per unit mass flow.

Combining (7.3.1) and (7.3.4) the *overall mechanical energy equation* may be written

$$\frac{p_3}{\rho} + \frac{v_3^2}{2} = \frac{p_1}{\rho} + \frac{v_1^2}{2} + W_i - (L_i + L_d) \tag{7.3.5}$$

or, in terms of head, the *actual gain of total head H* is given by

$$H = \left(\frac{p_3}{\rho g} + \frac{v_3^2}{2g}\right) - \left(\frac{p_1}{\rho g} + \frac{v_1^2}{2g}\right) = H_i - h_l, \qquad (7.3.6)$$

where h_l = total internal loss of head due to friction and eddying

$$= \frac{L_i + L_d}{g}.$$

The *hydraulic efficiency* is defined by $\eta_h = H/H_i$ or gH/W_i.

The *mechanical efficiency* is defined by $\eta_m = \dfrac{W_i}{\text{work input to pump}}$.

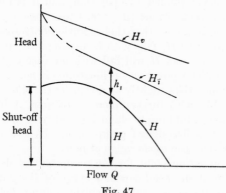

Fig. 47

Note that the work input to the pump includes the bearing losses and disk friction, etc., i.e. it is the total work done on the pump shaft by the driving motor.

The *overall efficiency* $\eta = \dfrac{gH}{\text{work input to pump}} = \eta_h \eta_m$.

7.4. The head-capacity curve

It is usual to plot the pump characteristic in the form of a curve of head H against volume flow Q for a given rotational speed. This is illustrated in fig. 47 together with a plot of the quantities H_i and H_v.

The difference, measured on the vertical scale, between the curve showing gain of total head H and the curve of head input H_i represents the total internal loss of head h_l caused by friction and eddying in the impeller and diffuser. The divergence between the curve of

H_i and the straight line giving H_v, on the other hand, does not represent a loss of energy but rather the difference between an actual head input and a hypothetical head input which could only be achieved if there was no secondary circulatory flow in the impeller. The variation of H_v with flow is given by equation (7.2.6) for the case of zero pre-rotation. Since the velocity f_2 is proportional to the volume rate of flow Q, the variation of H_v with Q will be linear. At shut-off with $f_2 = 0$, H_v would be equal to u_2^2/g, which is about twice the actual shut-off head.

It is evident from equation (7.2.6) that if the angle $\beta_2 < 90°$ H_v falls with increasing flow, but that if the angle $\beta_2 > 90°$ H_v would rise with increase of Q. A falling head-capacity curve is required in practice to achieve stability of operation, and it is for this reason that backward-swept blades ($\beta_2 < 90°$) are always employed. It should be noted, however, that this condition only ensures that the hypothetical quantity H_v falls with increasing flow, and it does not guarantee that the head H will fall with increase of flow over the whole range of Q. This possibility is illustrated in fig. 47, which shows a curve of H which rises slightly at first from the shut-off value and then falls steadily with further increase of flow. A head-capacity curve of this sort would imply slight instability of operation at low rates of flow. A well-designed pump should have a falling head-capacity curve over the *whole* range of flow.

The actual operating point for the pump will be determined by the intersection of the head-capacity curve of the pump with the characteristic of the system into which the pump is delivering.

7.5. Dimensional analysis applied to a pump

We can use the methods of Chapter 5 to analyse the performance of a range of pumps of similar geometrical design. We shall be concerned first of all with head H developed by the pump and its dependence on the rotational speed N, the impeller diameter D, the volumetric rate of flow Q, and the physical properties of the fluid. We can avoid introducing the gravitational acceleration g as a separate independent variable by taking the product gH as the dependent variable in this case. Reference to equation (7.2.4) or (7.3.6) will show that the head H always occurs in association with g and never by itself. We can therefore start by assuming that gH will be a function of (ρ, μ, N, D, Q), and by applying the technique of Chapter 5 this can be reduced to the dimensionless relationship

$$\frac{gH}{N^2 D^2} = f_1 \left(\frac{Q}{ND^3}, \frac{\rho ND^2}{\mu} \right). \tag{7.5.1}$$

Similarly for the power consumption P we can start by assuming that P will be a function of (ρ, μ, N, D, Q) and reduce this to

$$\frac{P}{\rho N^3 D^5} = f_2 \left(\frac{Q}{ND^3}, \frac{\rho ND^2}{\mu} \right). \tag{7.5.2}$$

Viscous effects are usually small compared with inertia effects owing to the high degree of turbulence and eddying inside a pump, and we

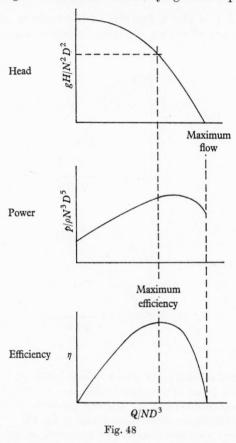

Fig. 48

can therefore say that in most cases (excluding the pumping of very viscous liquids)

$$\text{Head coefficient:} \quad c_H = \frac{gH}{N^2 D^2} = f_1 \left(\frac{Q}{ND^3} \right). \tag{7.5.3}$$

$$\text{Power coefficient:} \quad c_P = \frac{P}{\rho N^3 D^5} = f_2 \left(\frac{Q}{ND^3} \right). \tag{7.5.4}$$

Writing $c_Q = Q/ND^3$ for the flow coefficient, therefore, we should plot c_H against c_Q for the dimensionless head-capacity curve and c_P against c_Q for the power-capacity curve.

The overall efficiency is given by $\eta = \rho QgH/P$, which is equal to

$$\frac{Q}{ND^3}\frac{gH}{N^2D^2}\bigg/\frac{P}{\rho N^3D^5} \quad \text{or} \quad \frac{c_Q c_H}{c_P},$$

and therefore η is also a function of the capacity coefficient c_Q. Provided viscous effects are negligible, therefore, c_H, c_P and η can

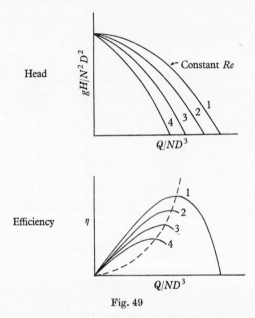

Fig. 49

each be plotted against c_Q to give a unique curve for a given pump design. At the point of maximum efficiency we will have definite values of these coefficients for a given geometrical design. These relationships are shown diagrammatically in fig. 48.

If, however, viscosity cannot be neglected as in the case of pumping a heavy oil or other viscous liquid, it is necessary to replace the single curve of c_H against c_Q by a family of curves each referring to a different value of the Reynolds number $\rho ND^2/\mu$. Similarly, a family of curves will be required for c_P and η. The effect of high viscosity or low values of the Reynolds number on the performance of a centrifugal pump is illustrated diagrammatically in this way in fig. 49. According to Stepanoff[1], the specific speed (see §7.6 below) at the maximum efficiency point remains the same with change of Reynolds number.

7.6. Specific speed of a pump

The specific speed N_s is defined as the speed in r.p.m. at which a geometrically similar pump (of reduced size) would develop a head of 1 ft. when delivering a flow of 1 gal./min. under similar operating conditions of maximum efficiency.

For similar operating conditions of maximum efficiency, we must fix the values of c_H and c_Q in the two cases at the point of maximum efficiency as indicated in fig. 48, i.e.

$$c_H = \frac{gH}{N^2 D^2} = \frac{g}{N_s^2 D_s^2} \quad \text{or} \quad \left(\frac{D_s}{D}\right)^2 = \frac{N^2}{N_s^2} \frac{1}{H},$$

where D_s is the reduced diameter of the pump which would operate at the specific speed N_s, and

$$c_Q = \frac{Q}{N D^3} = \frac{1}{N_s D_s^3} \quad \text{or} \quad \left(\frac{D_s}{D}\right)^3 = \frac{N}{N_s} \frac{1}{Q};$$

hence, eliminating D_s/D,

$$\frac{N}{N_s} \frac{1}{H^{\frac{1}{2}}} = \frac{N^{\frac{1}{3}}}{N_s^{\frac{1}{3}}} \frac{1}{Q^{\frac{1}{3}}},$$

i.e.

$$\frac{N^{\frac{2}{3}}}{N_s^{\frac{2}{3}}} = \frac{H^{\frac{1}{2}}}{Q^{\frac{1}{3}}},$$

or

$$N_s = \frac{N Q^{\frac{1}{2}}}{H^{\frac{3}{4}}}. \tag{7.6.1}$$

Note carefully that the specific speed defined in this way is a dimensional quantity and is usually measured with N in r.p.m., Q in g.p.m. and H in feet.

N_s is a 'type number' and its value gives an indication of the type of pump design which will be appropriate for operation at maximum efficiency while satisfying the required conditions of rotational speed, flow rate and head. It is particularly useful for classifying the range of application of different types of pump. The usual ranges are as follows:

	N_s
Single-stage radial flow centrifugal pumps	1,000–3,000
Mixed flow pumps	3,000–10,000
Axial flow pumps	10,000–15,000

These values are specified with N in r.p.m., Q in g.p.m. and H in feet.

For values of N_s below about 1000 it is generally necessary to employ a multi-stage centrifugal pump or alternatively to use some form of positive displacement pump.

It would be possible to define a *dimensionless specific speed* in the following way:

$$c_s = \frac{c_Q^{\frac{1}{2}}}{c_H^{\frac{3}{4}}} = \frac{NQ^{\frac{1}{2}}}{(gH)^{\frac{3}{4}}}. \tag{7.6.2}$$

In this case it is necessary to use a consistent set of units, e.g. N in radians/sec., Q in cusecs, H in feet, and g in ft./sec.[2] The definition given by (7.6.2) would be more sensible than that given by (7.6.1) for the specific speed, but the dimensionless form has not come into general use. The specific speed of a pump is usually quoted in the dimensional form of (7.6.1), and it is therefore necessary to check the units very carefully, particularly when comparing figures in British, Continental and American systems.

7.7. Cavitation

If the pressure in a fluid falls locally to a value below the vapour pressure at the prevailing temperature, cavities or bubbles are liable to be formed which will be filled with vapour. If such cavities are formed in a fluid stream, for instance, at the throat of a venturi tube or on the suction side of a pump impeller, the bubbles will move with the stream and will subsequently collapse when they reach a region of higher pressure. This phenomenon is known as cavitation. The collapse of the cavities leads to the formation of pressure waves in the fluid and these are liable to cause serious mechanical damage to the tube wall or pump impeller.

In the case of a centrifugal pump the onset of cavitation can be detected by the sudden increase in noise and vibration caused by the collapse of the vapour bubbles. It is also accompanied by a sudden fall in the head and efficiency curves when the flow is increased beyond the point at which cavitation starts. If a pump is left operating under cavitating conditions for any length of time the impeller will be subjected to mechanical damage in the form of pitting and apparent corrosion.

To avoid cavitation, attention must be paid to the design of the pump installation on the suction side. Referring to fig. 50, let p_0 be the atmospheric pressure on the suction side, let h_s be the height of the free surface on the suction side above the pump centre-line, and let h_f be the head lost due to friction, etc., in the suction line. The total head at the suction flange will then be given by

$$\left(\frac{p_0}{\rho g} + h_s - h_f\right).$$

The *net positive suction head* is defined as the difference between the total head at the suction flange and the head $p_v/\rho g$ corresponding to the absolute vapour pressure p_v at the prevailing temperature in the pump inlet, i.e.

$$\text{N.P.S.H.} = \left(\frac{p_0}{\rho g} + h_s - h_f\right) - \frac{p_v}{\rho g}. \qquad (7.7.1)$$

The average absolute pressure at entry to the pump impeller will be given by $p_0 + \rho g(h_s - h_f) - \dfrac{\rho v_1^2}{2}$, where v_1 is the average value of the absolute velocity at entry to the impeller. It will be reasonable to assume that there will be a further *local pressure drop* Δp below the average value at the eye of the impeller owing to non-uniformity

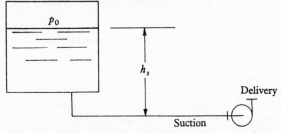

(Note that h_s will be numerically negative in the case of *suction lift*.)

Fig. 50

of the relative velocity of the fluid. We will assume further that this local pressure drop can be related to the average relative velocity at entry by the expression $\Delta p = \lambda\rho\,\dfrac{v_{r1}^2}{2}$, where λ is a pressure-drop coefficient. Under *critical conditions* for onset of cavitation, therefore, we have

$$p_0 + \rho g(h_s - h_f) - \rho\,\frac{v_1^2}{2} - \lambda\rho\,\frac{v_{r1}^2}{2} = p_v$$

or

$$\frac{p_0}{\rho g} + h_s - h_f = \frac{p_v}{\rho g} + \frac{v_1^2}{2g} + \lambda\frac{v_{r1}^2}{2g}, \qquad (7.7.2)$$

i.e.

$$\text{N.P.S.H.} = \left(\frac{p_0}{\rho g} + h_s - h_f\right) - \frac{p_v}{\rho g} = \frac{v_1^2}{2g} + \lambda\frac{v_{r1}^2}{2g}. \qquad (7.7.3)$$

To avoid cavitation, therefore, we want an N.P.S.H. greater than the maximum possible value of $\dfrac{v_1^2}{2g} + \lambda\dfrac{v_{r1}^2}{2g}$.

The coefficient λ should be a characteristic of the geometry of the pump design, and the values of the velocities v_1 and v_{r1} will depend on the flow rate Q. It should therefore be possible to estimate the minimum safe value of the net positive suction head from equation (7.7.3). A more simple procedure, however, has been suggested by Thoma who assumes that

$$\frac{v_1^2}{2g} + \lambda \frac{v_{r1}^2}{2g} = \sigma H, \qquad (7.7.4)$$

where σ is a *cavitation coefficient*, and H is the head developed by the pump. The value of the coefficient σ will depend on the type of pump, and in general it will be a function of the specific speed N_s. It is found experimentally that for single-entry pumps

$$\sigma = 0{\cdot}071 \ (N_s/1000)^{\frac{4}{3}} \text{ approximately.}$$

7.8. Numerical example of a pump calculation

A numerical example will illustrate the use of the various formulae derived in the preceding sections.

A single-stage centrifugal pump is designed to be driven at 1450 r.p.m. and to handle a flow of 1500 g.p.m. of a light oil of specific gravity 0·90. The pump has a single-suction impeller of 15 in. outside diameter, and the effective axial breadth between shrouds at the exit radius is 1 in. The impeller vanes are swept back at an angle of $22\frac{1}{2}°$ at exit. If the velocity of whirl at entry is neglected, calculate

(a) The actual head input by the impeller, assuming a mean velocity of whirl at exit equal to 0·7 of the ideal value.

(b) The head developed by the pump, assuming that internal losses amount to 60 per cent of the kinetic energy at outlet from the impeller.

(c) The power required to drive the pump if the equivalent head for the disk friction and bearing loss is $0{\cdot}12u^2/g$, where $u =$ peripheral velocity of the impeller.

(d) The maximum permissible lift (including friction loss) on the suction side if the cavitation coefficient σ is 0·133, and if the cavitation vapour pressure is 100 mm. Hg abs. at the prevailing temperature.

(a) $D = 1{\cdot}25$ ft. and $N = 1450$ r.p.m.

Peripheral velocity of the impeller $u_2 = 95$ ft./sec.

Flow area at exit from the impeller $= \pi D b = 0{\cdot}327$ sq.ft.

Volume flow $Q = 1500$ g.p.m. $= 4$ cusecs.

Flow velocity at exit from impeller $f_2 = 12 \cdot 2$ ft./sec.

Ideal whirl velocity $= u_2 - f_2 \cot \beta_2 = 65 \cdot 5$ ft./sec.

Actual whirl velocity is assumed to be $0 \cdot 7 \times$ ideal value, i.e. $w_2 = 45 \cdot 8$ ft./sec.

Head input $H_i = \dfrac{w_2 u_2}{g} = 135$ ft.

(b) Absolute velocity at exit from impeller $v_2 = \sqrt{(f_2^2 + w_2^2)} = 47 \cdot 4$ ft./sec.

Velocity head at exit from impeller $= v_2^2/2g = 34 \cdot 9$ ft.

Lost head assumed to be $0 \cdot 6\, v_2^2/2g$ in this case $= 21$ ft.

Actual head developed by the pump $= 135 - 21 = 114$ ft.

The *shut-off head* may be taken as $u_2^2/2g = 140$ ft.

(c) The equivalent head for the disk friction and bearing loss in the case considered is expressed as $0 \cdot 12 u_2^2/g = 33 \cdot 6$ ft.

Head equivalent of the shaft work $= H_i + 33 \cdot 6 = 168 \cdot 6$ ft.

Power required to drive the pump $= \dfrac{168 \cdot 6 \times 0 \cdot 9 \times 62 \cdot 5 \times 4}{550}$

$$= 69 \text{ h.p.}$$

(d) Specific speed $N_s = NQ^{\frac{1}{2}}/H^{\frac{3}{4}} = 1610$, which is within the range for a single-stage centrifugal pump.

Taking the cavitation coefficient as $0 \cdot 133$, the minimum value for the N.P.S.H. will be $0 \cdot 133 \times 114 = 15 \cdot 2$ ft.

But

$$\text{N.P.S.H.} = \frac{p_0}{\rho g} + h_s - h_f - \frac{p_v}{\rho g} = h_s - h_f + \frac{(p_0 - p_v)}{\rho g}.$$

$$\therefore \quad h_s - h_f = 15 \cdot 2 - \frac{660}{760} \times \frac{14 \cdot 7 \times 144}{0 \cdot 9 \times 62 \cdot 5}$$

$$= 15 \cdot 2 - 32 \cdot 7 = -17 \cdot 5 \text{ ft.},$$

i.e. maximum permissible suction lift $= 17 \cdot 5$ ft.

7.9. Classification of compressors

As in the case of pumps, compressors may be divided into the two main categories of *positive displacement machines* on the one hand and *turbo-machines* on the other.

Positive displacement compressors include both reciprocating machines and the rotary type. Turbo-compressors include both single- and multi-stage centrifugal compressors and also axial-flow machines. Generally speaking reciprocating compressors are used for compressing gases to high pressures, while turbo-compressors are used for handling larger rates of flow up to moderate pressure ratios.

The idealized case of reversible adiabatic compression of a *perfect gas* would give a relationship between pressure, volume and temperature according to the isentropic law

$$\frac{p_2}{p_1} = \left(\frac{V_1}{V_2}\right)^{\gamma} = \left(\frac{T_2}{T_1}\right)^{\gamma/(\gamma-1)} \tag{7.9.1}$$

but this cannot easily be achieved in practice.

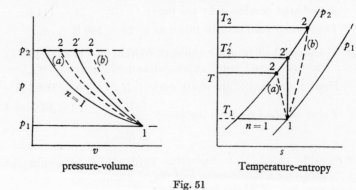

pressure-volume   Temperature-entropy

Fig. 51

The term *polytropic compression* is used in two different senses to describe compression according to the law

$$\frac{p_2}{p_1} = \left(\frac{V_1}{V_2}\right)^{n} = \left(\frac{T_2}{T_1}\right)^{n/(n-1)}, \tag{7.9.2}$$

where the exponent n differs in value from the ratio of the specific heats γ. Polytropic compression may mean either

(a) reversible but non-adiabatic compression, in which heat is supplied or removed during the process, or

(b) irreversible but adiabatic compression, in which there is a measurable thermal effect due to the action of friction and turbulence but no actual external transfer of heat.

Reciprocating compressors may approach case (a) provided velocities are small, and with cylinder cooling n will be less than γ.

Turbo-compressors approach case (b), i.e. friction effects are considerable but the flow is nearly adiabatic and n will be greater than γ.

The various possibilities are shown on the pressure-volume and temperature-entropy diagrams of fig. 51. If $n = 1$ we have the extreme case of *isothermal compression* with the relationship $pV = $ constant. This can never be realized in practice, since it would involve a process of slow reversible compression with the removal of exactly the right amount of heat $mRT \log_e V_1/V_2$ to maintain a constant temperature.

If $1 < n < \gamma$, we have the case represented by the dotted curve (a) shown on the p-V and T-s diagrams of fig. 51. If the process is reversible the work of compression per unit mass flow through the compressor can be evaluated from the integral $\int V \, dp$ and is given by

$$W = \frac{n}{n-1} (p_2 V_2 - p_1 V_1) \qquad (7.9.3)$$

or

$$W = \frac{nR}{n-1} (T_2 - T_1). \qquad (7.9.4)$$

Also from the steady-flow energy equation (3.3.2), neglecting changes of kinetic energy, we can say

$$W - Q = H_2 - H_1 = c_p (T_2 - T_1), \qquad (7.9.5)$$

where W is the shaft work done *on* the fluid and Q is the heat extracted *from* the fluid. From (7.9.4) and (7.9.5) the heat to be removed during the process, per unit mass flow, is given by

$$Q = \left[\frac{n}{n-1} - \frac{\gamma}{\gamma-1} \right] R(T_2 - T_1). \qquad (7.9.6)$$

The heat Q transferred during the process is represented graphically by the area $\int T \, ds$ under the curve 1-2 (a) on the temperature-entropy diagram of fig. 51 (the temperature T being measured from absolute zero).

If $n = \gamma$, and if there are no measurable effects from friction or eddying so that the process is thermodynamically reversible, we have the special case of *isentropic compression*. In this case the compression work W is

$$W = \frac{\gamma}{\gamma-1} (p_2 V_2 - p_1 V_1) = c_p (T_2' - T_1), \qquad (7.9.7)$$

where T_2' is given by

$$\frac{T_2'}{T_1} = \left(\frac{p_2}{p_1} \right)^{(\gamma-1)/\gamma}, \qquad (7.9.8)$$

and there is no transfer of heat.

If $n > \gamma$ we have the case represented by the dotted curve (b) in fig. 51. For a reversible process, heat would now have to be

8

supplied during the compression and the quantity would be represented graphically by the area $\int T \, ds$ under the curve 1-2(b) on the
temperature-entropy diagram, or numerically by

$$Q = \left[\frac{\gamma}{\gamma - 1} - \frac{n}{n - 1} \right] R(T_2 - T_1). \qquad (7.9.9)$$

If, on the other hand, we are considering case (b) with *irreversible*
but adiabatic compression, there is no external transfer of heat,
and the area beneath the curve on the T-s diagram or the quantity
calculated from (7.9.9) will now represent the *thermal equivalent of
the friction work*. With irreversible flow we can no longer say that the
shaft work is represented by $\int V \, dp$. Instead, we must use the steady-
flow energy equation (7.9.5) to calculate W and, since Q is now zero,

$$W = H_2 - H_1 = c_p(T_2 - T_1), \qquad (7.9.10)$$

the outlet temperature T_2 being given by

$$\frac{T_2}{T_1} = \left(\frac{p_2}{p_1} \right)^{(n-1)/n}. \qquad (7.9.11)$$

The flow through a turbo-compressor is always of a highly
irreversible nature with appreciable internal thermal effects due to
friction and turbulence. If there is no inter-stage cooling, however,
the process will be essentially adiabatic in the sense that no heat
will be supplied or rejected externally. The shaft work W will therefore be given by equation (7.9.10). It is usual to take the special case
of isentropic compression as a standard of comparison for the
performance of a turbo-compressor. The *isentropic efficiency* is
defined as the ratio of the work that would be required ideally for
isentropic compression between inlet pressure p_1 and delivery
pressure p_2 to the work actually required in the real process. From
(7.9.7) and (7.9.10), the isentropic efficiency η_c is given by

$$\eta_c = \frac{T_2' - T_1}{T_2 - T_1} = \frac{\left(\dfrac{p_2}{p_1} \right)^{(\gamma-1)/\gamma} - 1}{\dfrac{T_2}{T_1} - 1} \qquad (7.9.12)$$

or

$$\eta_c = \frac{\left(\dfrac{p_2}{p_1} \right)^{(\gamma-1)/\gamma} - 1}{\left(\dfrac{p_2}{p_1} \right)^{(n-1)/n} - 1}. \qquad (7.9.13)$$

For the ideal isentropic process we have $\dfrac{p}{p_1} = \left(\dfrac{T}{T_1}\right)^{\gamma/(\gamma-1)}$, and hence for a small change represented by the differentials $\mathrm{d}p$ and $\mathrm{d}T$

$$\frac{\mathrm{d}p}{p} = \frac{\gamma}{\gamma-1}\frac{\mathrm{d}T}{T}. \tag{7.9.14}$$

For a corresponding step with irreversible adiabatic compression we could say

$$\frac{\mathrm{d}p}{p} = \eta_p \frac{\gamma}{\gamma-1}\frac{\mathrm{d}T}{T}, \tag{7.9.15}$$

where η_p is the *polytropic efficiency*.

Alternatively for the actual irreversible process we could say

$$\frac{\mathrm{d}p}{p} = \frac{n}{n-1}\frac{\mathrm{d}T}{T}, \tag{7.9.16}$$

and hence, $\qquad \eta_p = \dfrac{n}{(n-1)}\dfrac{(\gamma-1)}{\gamma}. \tag{7.9.17}$

For a small pressure rise, for instance, with one stage of a multi-stage machine, the isentropic efficiency for the stage becomes identical with the polytropic efficiency defined by (7.9.17).

7.10. Flow through a centrifugal compressor

The mechanical arrangement of a centrifugal compressor is essentially similar to that of a centrifugal pump. Since higher velocities are normally required, however, impellers are frequently designed with straight radial blades because of the high centrifugal stresses. A single-stage radial flow compressor is represented diagrammatically in fig. 52 and velocity triangles are shown at inlet to and outlet from the impeller.

If there is no pre-rotation, the velocity of whirl w_1 at entry to the impeller should be zero. In this case, however, the relative velocity v_{r1} would have a tangential component equal to u_1 and the blades of the impeller should therefore be curved at entry. Alternatively, the compressor may be designed with fixed guide vanes to give pre-rotation so that the relative velocity v_{r1} may be nearly radial. With the latter arrangement w_1 will no longer be zero, but it will in any case be small compared with the velocity of whirl at exit and can generally be neglected in the calculations.

With straight radial blades, the relative velocity v_{r2} at exit from the impeller should also be radial. Owing to relative circulation and secondary flow effects, however, the relative velocity v_{r2} will in fact be inclined backwards at some angle β_2', as indicated in fig. 52. In other words the actual velocity of whirl at exit w_2 is less than the ideal value u_2, and we can say that $w_2 = \sigma u_2$, where σ is a *vane efficiency* or *slip factor*. Prediction of the appropriate value for σ in any particular case must generally be based on analysis of the performance of other compressors of similar design. A good indication of the value of the slip factor, however, is given by Stodola's formula $\sigma = (1 - \pi/n)$, where n is the number of blades.

The mechanics of the flow through the impeller is the same as that for the centrifugal pump, and we can make use of the angular

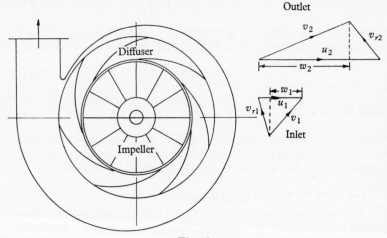

Fig. 52

momentum equation to derive the following expression for the work done on the fluid by the impeller per unit mass flow:

$$W_i = (w_2 u_2 - w_1 u_1). \tag{7.10.1}$$

For the case of radial blades $w_2 = \sigma u_2$ and therefore

$$W_i = \sigma u_2^2 - w_1 u_1$$

or $\quad W_i = \sigma u_2^2 \quad$ approximately if $w_1 u_1$ is small (cf. u_2^2).

The energy equation for flow through the impeller may be written as

$$H_2 + \frac{v_2^2}{2} = H_1 + \frac{v_1^2}{2} + W_i. \tag{7.10.2}$$

Note that we are using the *complete* energy equation here so that there is no term for mechanical energy loss. If the fluid is a gas whose enthalpy can be expressed as $H = Jc_pT$, the equation becomes

$$Jc_pT_2 + \frac{v_2^2}{2} = Jc_pT_1 + \frac{v_1^2}{2} + W_i. \tag{7.10.3}$$

The energy equation for flow through the diffuser is

$$H_3 + \frac{v_3^2}{2} = H_2 + \frac{v_2^2}{2}$$

or
$$Jc_pT_3 + \frac{v_3^2}{2} = Jc_pT_2 + \frac{u_2^2}{2}, \tag{7.10.4}$$

where T_3 and v_3 are the temperature and velocity at exit from the diffuser.

Similarly for flow up to the eye of the impeller, the energy equation is

$$Jc_pT_1 + \frac{v_1^2}{2} = Jc_pT_0 + \frac{v_0^2}{2}, \tag{7.10.5}$$

where T_0 and v_0 are the temperature and velocity at entry to the machine (i.e. at the suction flange).

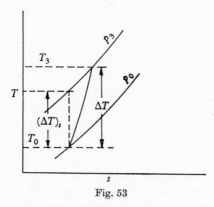

Fig. 53

The overall energy equation for flow through the machine is simply the sum of (7.10.3), (7.10.4) and (7.10.5), i.e.

$$Jc_pT_3 + \frac{v_3^2}{2} = Jc_pT_0 + \frac{v_0^2}{2} + W_i. \tag{7.10.6}$$

Note that the mechanical equivalent of heat J has been introduced in the enthalpy terms in these equations so that we are working in terms of mechanical units throughout.

If the velocity v_3 at the delivery flange is approximately equal to the velocity v_0 at the suction flange, the last equation is simplified to

$$Jc_p(T_3 - T_0) = W_i$$

or
$$\Delta T = T_3 - T_0 = \frac{\sigma u_2^2 - w_1 u_1}{Jc_p}. \tag{7.10.7}$$

Referring to the temperature-entropy diagram of fig. 53, the actual temperature rise ΔT can be related to the hypothetical temperature rise $(\Delta T)_s$ for isentropic compression by

$$(\Delta T)_s = \eta_c \Delta T, \tag{7.10.8}$$

where η_c is the isentropic efficiency. The overall pressure ratio is therefore given by

$$\frac{p_3}{p_0} = \left[1 + \frac{\eta_c \Delta T}{T_0}\right]^{\gamma/(\gamma-1)} = \left[1 + \frac{\eta_c(\sigma u_2^2 - w_1 u_1)}{Jc_p T_0}\right]^{\gamma/(\gamma-1)} \tag{7.10.9}$$

or, alternatively, by

$$\frac{p_3}{p_0} = \left[1 + \frac{\Delta T}{T_0}\right]^{n/(n-1)} = \left[1 + \frac{(\sigma u_2^2 - w_1 u_1)}{Jc_p T_0}\right]^{n/(n-1)}. \tag{7.10.10}$$

7.11. Dimensional analysis applied to a turbo-compressor

We must now extend the use of dimensional analysis by introducing another fundamental dimension Θ for temperature. A full discussion of the use of dimensional analysis in problems involving thermal effects will be given in Chapter 10, but for the present we will simply analyse the performance of a turbo-compressor using the same procedure as in §§5.5 and 7.5 but using the four fundamental dimensions M, L, T, Θ.

We start by assuming that for a range of geometrically similar machines the delivery pressure p will be a function of the inlet pressure and temperature, p_0 and T_0, the physical properties of the gas c_p and γ, the rotational speed N, the impeller diameter D, and the mass flow m. We can omit the viscosity of the gas, since the flow will be extremely turbulent and inertia forces will completely outweigh viscous forces. The starting point therefore is the statement that

$$p = f(p_0,\ T_0,\ c_p,\ \gamma,\ N,\ D,\ m). \tag{7.11.1}$$

There are eight variables and four dimensions and the statement can therefore be reduced to a relationship between four dimensionless ratios. Taking p_0, T_0, c_p and D as the primary quantities, and using the method of Chapter 5, we find

$$\frac{p}{p_0} = f\left[\frac{m\sqrt{(c_p T_0)}}{D^2 p_0},\ \frac{ND}{\sqrt{(c_p T_0)}},\ \gamma\right]. \tag{7.11.2}$$

If we are dealing with gases having the same value for the ratio γ, we can plot the pressure ratio p/p_0 against the dimensionless mass-flow coefficient $m\sqrt{(c_p T_0)}/D^2 p_0$ for various constant values of the dimensionless speed coefficient $ND/\sqrt{(c_pT_0)}$. This is indicated in fig. 54 and represents the characteristic for a turbo-compressor equivalent to the dimensionless plot of the head-capacity curve for a pump.

Unlike a centrifugal pump, however, a turbo-compressor can only operate over a limited part of the pressure ratio-mass flow characteristic. For every value of the ratio $ND/\sqrt{(c_pT_0)}$ there is a minimum value for the ratio $m\sqrt{(c_pT_0)}/D^2p_0$ below which the flow becomes

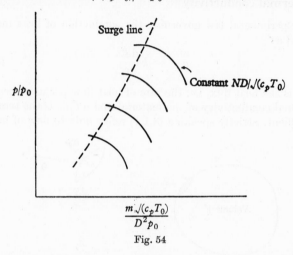

Fig. 54

unstable due to stalling of the diffuser guide vanes and separation of flow inside the compressor. This is indicated by the *surge line* in fig. 54. There is also a top limit to the mass flow or rather to the ratio $m\sqrt{(c_pT_0)}/D^2p_0$, for any value of $ND/\sqrt{(c_pT_0)}$, where the pressure ratio begins to fall sharply and the characteristic curves become vertical.

REFERENCE

(1)   STEPANOFF. *Centrifugal and Axial Flow Pumps* (Wiley).

CHAPTER 8

HEAT CONDUCTION AND HEAT TRANSFER

8.1. Thermal conductivity

The experimental law governing the conduction of heat may be expressed as

$$q = - k \frac{\mathrm{d}T}{\mathrm{d}r},\qquad(8.1.1)$$

where q is the heat flux, i.e. the rate of heat flow per unit area, k is the thermal conductivity of the material and $\mathrm{d}T/\mathrm{d}r$ is the temperature gradient. Strictly speaking (8.1.1) refers only to flow of heat in

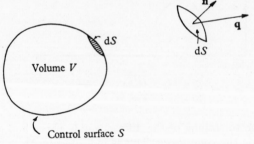

Fig. 55

the direction of the radius r, for instance, through the wall of a tube. The general form of the conduction law is expressed in vector notation as follows

$$\mathbf{q} = - k \operatorname{grad} T.\qquad(8.1.2)$$

The thermal conductivity k is usually treated as a constant for any given material, but it does in fact vary appreciably with temperature in some cases. Note that numerical values for the thermal conductivity are usually quoted in B.TH.U./ft.hr.°F.

8.2. Continuity of heat flow in a solid

The mathematical equation for continuity in heat flow is derived in a similar way to the equation of continuity for mass flow given in §2.2. Consider a control surface S enclosing a volume V as shown in

104

fig. 55. The rate of accumulation of internal energy inside the surface S is given by

$$\frac{\partial}{\partial t} \oint_V \rho c T \, \mathrm{d}V,$$

where c is the specific heat of the solid. The net rate of outflow of heat across the surface S is given by

$$\oint_S \mathbf{q} \cdot \mathbf{n} \, \mathrm{d}S = \oint_V \mathrm{div} \, \mathbf{q} \, \mathrm{d}V,$$

where $\mathbf{q}$ is the heat flux or current density. Conservation of energy, therefore, requires that

$$\frac{\partial}{\partial t} \oint_V \rho c T \, \mathrm{d}V + \oint_V \mathrm{div} \, \mathbf{q} \, \mathrm{d}V = 0$$

or

$$\rho c \frac{\partial T}{\partial t} + \mathrm{div} \, \mathbf{q} = 0. \qquad (8.2.1)$$

Substituting for $\mathbf{q}$ from (8.1.2) gives

$$\rho c \frac{\partial T}{\partial t} - \mathrm{div} \, (k \, \mathrm{grad} \, T) = 0, \qquad (8.2.2)$$

and if k is constant,

$$\rho c \frac{\partial T}{\partial t} = k \, \mathrm{div} \, \mathrm{grad} \, T \qquad (8.2.3)$$

or

$$\frac{\partial T}{\partial t} = \frac{k}{\rho c} \nabla^2 T, \qquad (8.2.4)$$

which is the Fourier heat conduction equation. $k/\rho c$ is the *thermal diffusivity*, α having dimensions L^2/T.

If heat is being generated internally as a result of the passage of an electric current or of some process of radioactive decay or other nuclear reaction, at local rate H per unit volume, equation (8.2.3) becomes

$$\rho c \frac{\partial T}{\partial t} = k \nabla^2 T + H. \qquad (8.2.5)$$

H is not necessarily a constant and may in fact be a function of position.

The mathematical treatment of heat conduction consists in finding solutions to the Fourier conduction equation which will satisfy certain initial and boundary conditions. As an example of the procedure, consider the problem of the flow of heat in a large slab or flat plate. We will assume that the surface dimensions are large compared with the thickness and that the temperature at any instant is uniform over each surface. The problem may therefore be treated as one of linear flow of heat varying with time. Referring to fig. 56 the equation to be solved is

$$\frac{\partial T}{\partial t} = \frac{k}{\rho c}\frac{\partial^2 T}{\partial x^2},$$ (8.2.6)

which is simply a special case of (8.2.4).

The *initial conditions* will specify the temperature distribution at the instant $t = 0$, i.e. the solution must satisfy the condition that $T = f(x)$ when $t = 0$.

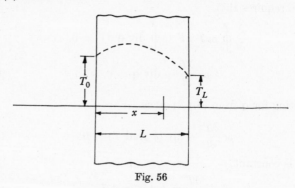

Fig. 56

The *boundary conditions* at the surfaces will depend on the particular problem considered, but the conditions most commonly encountered are:

(i) surface temperature constant, i.e. $T = T_0$ at $x = 0$, and $T = T_L$ at $x = L$, or

(ii) thermal contact with some other conducting medium, requiring continuity of heat flux at the surface, i.e. $k_a\dfrac{\partial T_a}{\partial x} = k_b\dfrac{\partial T_b}{\partial x}$, where a and b refer to the two different materials on either side of the boundary, or

(iii) heat flux at the surface determined by radiation or convection to a surrounding gas or fluid, in which case, at the surface $x = 0$ for instance, $q_0 = -k(\partial T/\partial x)_0 = h(T_0 - \theta_0)$, where h is the heat-transfer coefficient and θ_0 is the temperature of the surrounding fluid (this condition is sometimes referred to as Newton's law of cooling), or

(iv) complete thermal insulation at a surface, e.g. at the surface $x = 0$ with zero heat flux $(\partial T/\partial x)_0 = 0$.

For the various solutions of the partial differential equation

(8.2.6) which will satisfy these different boundary conditions the reader should consult [1] or [2].

Other special cases of equation (8.2.4) for which solutions are derived in the textbooks dealing with the mathematical theory of heat conduction are as follows:

Radial flow of heat in a cylinder:

$$\frac{\partial T}{\partial t} = \frac{k}{\rho c} \left[\frac{\partial^2 T}{\partial r^2} + \frac{1}{r} \frac{\partial T}{\partial r} \right]. \tag{8.2.7}$$

Radial flow of heat in a sphere:

$$\frac{\partial T}{\partial t} = \frac{k}{\rho c} \left[\frac{\partial^2 T}{\partial r^2} + \frac{2}{r} \frac{\partial T}{\partial r} \right]. \tag{8.2.8}$$

Two-dimensional flow of heat in a plate:

$$\frac{\partial T}{\partial t} = \frac{k}{\rho c} \left[\frac{\partial^2 T}{\partial x^2} + \frac{\partial^2 T}{\partial y^2} \right]. \tag{8.2.9}$$

Radial and axial flow of heat in a cylinder:

$$\frac{\partial T}{\partial t} = \frac{k}{\rho c} \left[\frac{\partial^2 T}{\partial r^2} + \frac{1}{r} \frac{\partial T}{\partial r} + \frac{\partial^2 T}{\partial z^2} \right]. \tag{8.2.10}$$

Three-dimensional flow of heat in a slab:

$$\frac{\partial T}{\partial t} = \frac{k}{\rho c} \left[\frac{\partial^2 T}{\partial x^2} + \frac{\partial^2 T}{\partial y^2} + \frac{\partial^2 T}{\partial z^2} \right]. \tag{8.2.11}$$

Note that these all refer to unsteady heat flow, i.e. heat conduction with a temperature distribution which varies with time.

8.3. Steady flow of heat in one dimension

The problem of heat conduction is greatly simplified if we are only concerned with *steady conditions*. The general equation in this case follows from (8.2.2)

$$\text{div} \, (k \, \text{grad} \, T) = 0 \tag{8.3.1}$$

or, with a *constant value for the thermal conductivity,*

$$\nabla^2 T = 0. \tag{8.3.2}$$

For steady conduction through a wall or slab, treated as a one-dimensional problem, equation (8.3.2) becomes

$$\frac{\mathrm{d}^2 T}{\mathrm{d}x^2} = 0, \tag{8.3.3}$$

and the solution is

$$(T_1 - T) = (T_1 - T_2)\frac{x}{L},\qquad(8.3.4)$$

i.e. a linear temperature distribution. The heat flux is

$$q = -k\frac{dT}{dx} = k\frac{(T_1 - T_2)}{L}$$

or, for a cross-sectional area A, the total flow of heat is given by

$$Q = qA = \frac{kA}{L}(T_1 - T_2) = \frac{T_1 - T_2}{R},\qquad(8.3.5)$$

where R is the thermal resistance to heat flow.

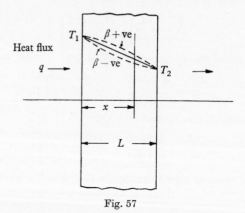

Fig. 57

With many materials the thermal conductivity k is not constant but varies in a nearly linear manner with temperature, i.e. we can say approximately that

$$k = k_0[1 + \beta(T - T_0)],\qquad(8.3.6)$$

where β is a temperature coefficient. For most metals β is negative, although aluminium and certain non-ferrous alloys are exceptions to the rule. For most non-metals β is positive. With a *variable conductivity* we must go back to equation (8.3.1) which becomes for the one-dimensional problem

$$\frac{d}{dx}\left(k\frac{dT}{dx}\right) = 0.\qquad(8.3.7)$$

Substituting for k from (8.3.6) and integrating from $x = 0$ to $x = L$ gives

$$- k_0 \left[1 + \beta \left(\frac{T_2 + T_1}{2} - T_0 \right) \right] (T_2 - T_1) = qL$$

or

$$- k_m(T_2 - T_1) = qL,$$

i.e.   heat flux $q = \dfrac{k_m}{L}(T_1 - T_2),$ (8.3.8)

where

$$k_m = k_0 \left[1 + \beta \left(\frac{T_2 + T_1}{2} - T_0 \right) \right],$$

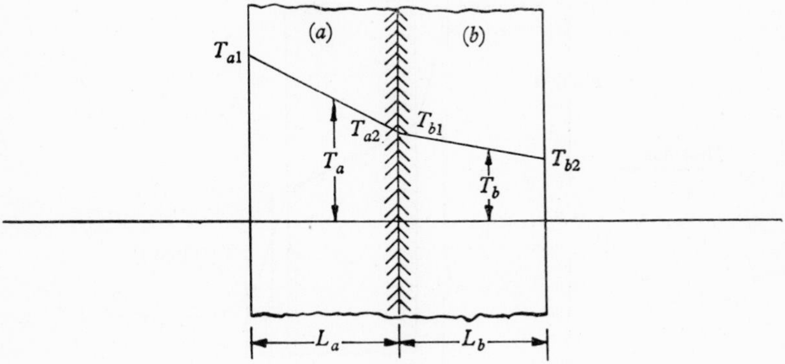

Fig. 58

which is the thermal conductivity evaluated at the arithmetic mean temperature. The temperature distribution, however, will no longer be linear but will be given by

$$(T_1 - T)(1 - \beta T_0) + \frac{\beta}{2}(T_1^2 - T^2) = \frac{qx}{k_0}. \tag{8.3.9}$$

The two cases of β positive and β negative are shown by the dotted lines on fig. 57.

Reverting to the case of constant thermal conductivity, the state of affairs at a *change of medium* with complete thermal contact is shown in fig. 58. The boundary conditions are:

For complete contact: $T_{a_2} = T_{b_1}.$

For continuity of heat flow: $k_a \dfrac{\mathrm{d}T_a}{\mathrm{d}x} = k_b \dfrac{\mathrm{d}T_b}{\mathrm{d}x}.$

Using the notation of fig. 58

$$T_{a_1} - T_{a_2} = q\,\frac{L_a}{k_a},$$

$$T_{b_1} - T_{b_2} = q\,\frac{L_b}{k_b},$$

hence, $$T_{a_1} - T_{b_2} = q\left(\frac{L_a}{k_a} + \frac{L_b}{k_b}\right)$$

or $$q = \frac{T_{a_1} - T_{b_2}}{(L_a/k_a + L_b/k_b)}. \qquad (8.3.10)$$

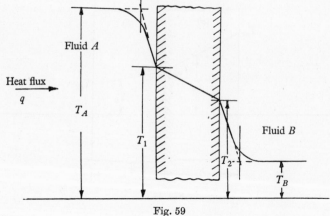

Fig. 59

Similarly for *series flow* through several layers of material, if the temperatures at the boundaries are $T_1, T_2, T_3, \ldots, T_n$,

$$T_1 - T_n = q\Sigma\left(\frac{L}{k}\right),$$

or $$q = \frac{T_1 - T_n}{\Sigma(L/k)}. \qquad (8.3.11)$$

For *conduction through a solid wall from one fluid to another* as indicated in fig. 59 we have the following conditions:

For heat transfer at surface (1): $q = h_1(T_A - T_1)$.

For conduction through the wall: $q = \dfrac{k}{L}\,(T_1 - T_2)$.

For heat transfer at surface (2): $q = h_2(T_2 - T_B)$.

Note that the expression for the heat transfer at a surface is really a definition of a *heat-transfer factor* or coefficient h. For continuity of heat flux, it follows that

$$T_A - T_B = q \left(\frac{1}{h_1} + \frac{L}{k} + \frac{1}{h_2} \right)$$

or

$$q = \frac{T_A - T_B}{\left(\frac{1}{h_1} + \frac{L}{k} + \frac{1}{h_2} \right)}. \qquad (8.3.12)$$

8.4. Steady radial flow of heat in a cylinder

For steady radial flow of heat in a long cylinder, equation (8.3.2) takes the form

$$\frac{d^2 T}{dr^2} + \frac{1}{r} \frac{dT}{dr} = 0. \qquad (8.4.1)$$

This also follows directly from equation (8.2.7).

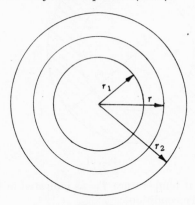

Fig. 60

Integrating equation (8.4.1) we get

$$r \frac{dT}{dr} = \text{constant}, \qquad (8.4.2)$$

and this is simply the condition that the total flow of heat per unit length is constant, i.e.

$$Q = -2\pi r \, k \frac{dT}{dr}.$$

Integrating (8.4.2) and putting in the boundary conditions that

$T = T_1$ at $r = r_1$ and $T = T_2$ at $r = r_2$ gives the temperature distribution

$$(T_1 - T) = (T_1 - T_2)\frac{\log_e r/r_1}{\log_e r_2/r_1}, \qquad (8.4.3)$$

and the heat flux q is given by

$$q = -k\frac{\mathrm{d}T}{\mathrm{d}r} = \frac{k}{r}\frac{(T_1 - T_2)}{\log_e r_2/r_1}. \qquad (8.4.4)$$

The total flow of heat Q per unit length is

$$Q = 2\pi r q = \frac{2\pi k(T_1 - T_2)}{\log_e r_2/r_1}. \qquad (8.4.5)$$

For *heat transfer through the wall of a circular pipe*, from a fluid (A) flowing inside the pipe at temperature T_A, to another fluid (B)

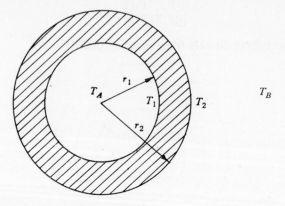

Fig. 61

outside the pipe at temperature T_B, as indicated in fig. 61, we have the following three conditions:

Heat transfer at inner surface (1) $Q = 2\pi r_1 h_1(T_A - T_1)$.

Conduction through the wall $Q = 2\pi k\dfrac{(T_1 - T_2)}{\log_e r_2/r_1}$.

Heat transfer at outer surface (2) $Q = 2\pi r_2 h_2(T_2 - T_B)$.

Hence, $$Q = \frac{2\pi(T_A - T_B)}{\left(\dfrac{1}{r_1 h_1} + \dfrac{1}{k}\log_e\dfrac{r_2}{r_1} + \dfrac{1}{r_2 h_2}\right)}. \qquad (8.4.6)$$

Another example of radial heat flow arises with the lagging of a steam pipe with material of low thermal conductivity to reduce the rate of heat loss. Let a be the outer radius of the pipe with surface temperature T_s and let b be the outer radius of the lagging. The rate of heat loss is then given by

$$Q = \frac{2\pi(T_s - T_0)}{\frac{1}{k_L}\log_e\frac{b}{a} + \frac{1}{bh}}, \qquad (8.4.7)$$

where k_L is the thermal conductivity of the lagging, h is the heat-transfer coefficient for free convection at the outer surface, T_0 is the temperature of the surrounding atmosphere.

For heat flow in a solid cylinder with *internal generation of heat*, at a uniform rate H per unit volume of material, the radial-flow conduction equation is

$$k\left[\frac{\mathrm{d}^2T}{\mathrm{d}r^2} + \frac{1}{r}\frac{\mathrm{d}T}{\mathrm{d}r}\right] + H = 0 \qquad (8.4.8)$$

or

$$\frac{\mathrm{d}}{\mathrm{d}r}\left[r\frac{\mathrm{d}T}{\mathrm{d}r}\right] = -\frac{Hr}{k},$$

hence, integrating the last equation from 0 to r, we get

$$r\frac{\mathrm{d}T}{\mathrm{d}r} = -\frac{Hr^2}{2k}$$

or

$$\frac{\mathrm{d}T}{\mathrm{d}r} = -\frac{Hr}{2k}.$$

Integrating again from 0 to r, we get the temperature distribution

$$T_0 - T = \frac{Hr^2}{4k}, \qquad (8.4.9)$$

where T_0 is the temperature at the centre of the cylinder. The total outward flow of heat at radius r is given by

$$Q = -2\pi rk\frac{\mathrm{d}T}{\mathrm{d}r} = \pi r^2H. \qquad (8.4.10)$$

8.5. Steady flow of heat in a thin rod or fin

Consider the steady flow of heat along a thin rod or fin as indicated in fig. 62. Let the temperature at the base of the fin be T_1, let the temperature at distance x along the fin be T, and let the temperature of the surrounding atmosphere be T_0. We will assume that heat is

9

being transferred from the surface of the fin by convection and that the loss of heat per unit area of surface can be expressed as

$$q = h(T - T_0). \tag{8.5.1}$$

We will make the further simplifying assumption that the local heat transfer factor h is constant over the entire surface of the fin. Let the area of cross-section of the fin be a, and the perimeter p. We will neglect any temperature variation over the cross-section and will assume that $T = f(x)$ only.

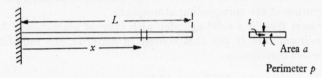

Fig. 62

Considering an element of the fin of length dx as shown in fig. 62:

Net outflow of heat by conduction $= \dfrac{dQ}{dx}\,dx = -\,ak\,\dfrac{d^2T}{dx^2}\,dx.$

Outflow of heat by convection from the surface $= h(T - T_0)p\,dx.$

Hence for continuity

$$ak\,\frac{d^2T}{dx^2}\,dx - h(T - T_0)p\,dx = 0$$

or
$$\frac{d^2T}{dx^2} - \frac{hp}{ak}(T - T_0) = 0. \tag{8.5.2}$$

Substituting $\theta = T - T_0$, equation (8.5.2) may be written

$$\frac{d^2\theta}{dx^2} - \lambda^2\theta = 0, \tag{8.5.3}$$

where $\lambda = \sqrt{(hp/ak)}$. For a *wide fin*, taking unit length perpendicular to the axis and neglecting the contribution of the thickness t to the perimeter, the ratio $\lambda = \sqrt{(2h/kt)}$.

The solution to equation (8.5.3) is

$$\theta = A\cosh\lambda x + B\sinh\lambda x, \tag{8.5.4}$$

and the boundary conditions are

at $x = 0$: $\theta = \theta_1$, $\therefore A = \theta_1$;

at $x = L$: $\dfrac{d\theta}{dx} = 0$ (neglecting heat transfer from the end of the fin);

i.e. $\lambda A \sinh \lambda L + \lambda B \cosh \lambda L = 0$,

therefore $B = -\theta_1 \dfrac{\sinh \lambda L}{\cosh \lambda L}$.

The temperature distribution is then given by

$$\theta = \theta_1 \cosh \lambda x - \theta_1 \frac{\sinh \lambda L}{\cosh \lambda L} \sinh \lambda x$$

or $$\theta = \theta_1 \frac{\cosh \lambda (L - x)}{\cosh \lambda L}. \tag{8.5.5}$$

The flow of heat into the rod or fin at $x = 0$ is given by

$$Q_0 = -ka \left(\frac{d\theta}{dx}\right)_{x=0} = \lambda ka \theta_1 \left[\frac{\sinh \lambda (L - x)}{\cosh \lambda L}\right]_{x=0},$$

i.e. $$Q_0 = \lambda ka \theta_1 \tanh \lambda L, \tag{8.5.6}$$

and this is equal to the lateral loss of heat by convection or radiation.

For a wide fin, taking unit length perpendicular to the x-axis, the total flow of heat into the base of the fin is given by

$$Q_0 = \lambda kt \theta_1 \tanh \lambda L. \tag{8.5.7}$$

The *effectiveness of a fin* is defined as the ratio of the heat actually transferred through the fin to the heat that would be transferred by convection or radiation from the base area if the fin was removed. It is assumed in this definition that the local heat-transfer factor for convection from the base area would be the same as the average value of the heat-transfer factor over the surface of the fin. From (8.5.7) the effectiveness is given by

$$\frac{\lambda kt \theta_1 \tanh \lambda L}{ht \theta_1}$$

or, since $\lambda^2 = 2h/kt$, the effectiveness $= \dfrac{2}{\lambda t} \tanh \lambda L$.

8.6. Extended surfaces for heat transfer

In the design of heat exchangers, extended surfaces are employed in cases where the heat-transfer factor is particularly low and a very large surface area is required. For instance, in a gas to liquid heat exchanger employing a tube bank, with the liquid flowing inside the tubes and the gas flowing across the tubes on the outside, the main resistance to heat transfer will generally occur on the gas side. It may therefore be worth while to use tubes having fins or some other form of extended surface such as studs or pins to provide additional surface area on the gas side.

It is generally advisable to base any design calculations on actual experimental results with the particular type of extended surface proposed. This is largely on account of the unpredictable variation

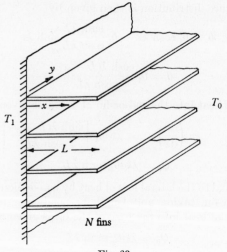

Fig. 63

in the value of the local heat-transfer factor. The following simple theory for a flat surface with plane fins, however, will explain the principles involved.

Referring to fig. 63, let the area of the flat surface be A and let the temperature be T_1. Let there be N fins of length L and thickness t. Consider unit length in the y direction.

Area of unfinned surface $= A$.

Area of finned surface $= A_f = 2LN$.

Rate of heat flow from unfinned surface $= Ah(T_1 - T_0)$.

Rate of heat flow from the finned surface, from equation (8.5.7),

$$= N\lambda kt(T_1 - T_0) \tanh \lambda L$$

$$= \frac{A_f h_f}{\lambda L}(T_1 - T_0) \tanh \lambda L.$$

The total flow of heat is therefore given by

$$Q = \left[Ah + A_f h_f \frac{1}{\lambda L} \tanh \lambda L \right] (T_1 - T_0). \qquad (8.6.1)$$

Note that the average heat-transfer factor h_f for the finned surface is not necessarily the same as the local heat-transfer factor h for the flat surface or wall.

The quantity $\dfrac{1}{\lambda L} \tanh \lambda L$ which occurs in (8.6.1) is the *fin efficiency*. It is numerically always less than 1·0 and is the ratio by which $A_f h_f (T_1 - T_0)$ must be multiplied to take account of the fact that the full temperature difference $T_1 - T_0$ is not available for heat transfer from the surface of the fins. Note that λ is defined as being equal to $\sqrt{(2h_f/kt)}$.

Other problems on finned surfaces include circular pipes with radial or helical fins, and flat surfaces with tapered fins. These may be investigated by methods similar to those of §8.5 but involve greater mathematical complexity. For further information on extended surfaces the reader should consult [3].

REFERENCES

(1)   CARSLAW and JAEGER. *Conduction of Heat in Solids* (Oxford).
(2)   INGERSOLL, ZOBEL and INGERSOLL. *Heat Conduction* (McGraw-Hill).
(3)   KERN. *Process Heat Transfer* (McGraw-Hill).

Chapter 9

HEAT EXCHANGERS

9.1. Heat-transfer factors

The starting point for any design calculation on a heat exchanger is the definition of a heat-transfer factor h

$$q = h\Delta T, \tag{9.1.1}$$

where q is the heat flux per unit area of surface and ΔT is the temperature difference between the surface and the fluid at the point considered. Values of h are usually quoted in B.TH.U./sq. ft.hr.°F.

If we are content to use an average temperature difference ΔT_m and an average value for the heat-transfer factor h, the total rate of heat transfer for a heat exchanger of surface area A will be given by

$$Q = hA\Delta T_m. \tag{9.1.2}$$

It is very seldom that the local variation in the value of the heat-transfer factor is known with any accuracy. Experimental values are usually obtained from measured rates of heat transfer with complete heat exchangers and are therefore necessarily mean values. The use of dimensional analysis to provide a logical method of correlating such experimental values will be outlined in Chapter 10. The following table of typical values for h, however, will give an idea of the orders of magnitude involved with different types of heat exchanger:

Heat-transfer operation	Usual range for h (B.TH.U./sq.ft. hr. °F.)
Gas flow across tube banks	2 to 10
Liquids flowing across tubes	10 to 100
Organic liquids inside tubes	50 to 500
Water flowing inside tubes	100 to 1,000
Boiling liquids	200 to 2,000
Condensing vapours	300 to 3,000

If we are concerned with heat transfer from one fluid to another three separate transfer operations are involved: convective transfer from the first fluid to the wall or surface of the heat exchanger, conduction through the wall, and convective transfer to the second fluid. This has already been mentioned in §8.3, and it follows from

118

equation (8.3.12) that the *overall heat-transfer factor* h_0 is determined, in the case of flat surfaces, by the following relationship:

$$\frac{1}{h_0} = \frac{1}{h_1} + \frac{L}{k} + \frac{1}{h_2}, \tag{9.1.3}$$

where h_1 and h_2 are the individual factors for the convective transfer operation on either side, L is the length of the conduction path and k is the thermal conductivity of the material of the heat exchanger wall.

For a *tubular heat exchanger*, from equation (8.4.6), if the overall heat-transfer factor is referred to the *inner surface area*

$$\frac{1}{h_0} = \frac{1}{h_1} + \frac{r_1}{k} \log_e \frac{r_2}{r_1} + \frac{r_1}{r_2 h_2}, \tag{9.1.4}$$

and if the overall heat-transfer factor is referred to the *outer surface area*

$$\frac{1}{h_0} = \frac{r_2}{r_1 h_1} + \frac{r_2}{k} \log_e \frac{r_2}{r_1} + \frac{1}{h_2}. \tag{9.1.5}$$

In addition to the three individual resistances to heat transfer already mentioned, there may be an additional resistance associated with the formation of a film of dirt or scale deposit on the surface. This is usually allowed for by including a *fouling factor* when calculating the overall heat-transfer factor. In place of (9.1.3), for instance, we must have

$$\frac{1}{h_0} = \frac{1}{h_1} + \frac{L}{k} + \frac{1}{h_2} + R, \tag{9.1.6}$$

where R is the fouling factor measured in ft.^{2} hr.°F./B.TH.U. Typical values for R are as follows:

	Usual range for R (sq.ft. hr. °F./B.TH.U.)
Fuel oil	0·004 to 0·010
Light oils and other organic liquids	0·001 to 0·004
River water	0·002 to 0·003
Boiler feed water (treated)	0·0005 to 0·001

Further discussion of the calculation of heat-transfer factors is reserved for Chapter 10.

9.2. Mean temperature difference for flow in a pipe with the wall at uniform temperature

In order to make use of equation (9.1.2) we require to know the correct value for the mean temperature difference ΔT_m. We will consider first of all the case of a tube with the wall maintained at a uniform temperature and a single-phase fluid flowing inside the tube. Referring to fig. 64, let the fluid enter at $x = 0$ with temperature T_1 and let it leave at $x = L$ with temperature T_2. Let the temperature of the tube wall be T_0. Both the cases of cooling and of heating of the fluid are shown on fig. 64. In the following argument,

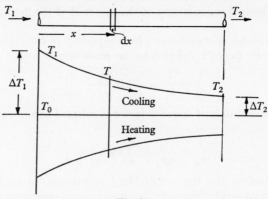

Fig. 64

however, we will deal with cooling of the fluid. The final result is the same in either case. Taking a *heat balance for an element* dx,

Heat flow to the pipe wall in length $dx = q\pi D \, dx$
$$= h\pi D(T - T_0) \, dx,$$

Heat given up by the fluid in length $dx = -\rho u_m c_p \dfrac{\pi D^2}{4} \, dT,$

hence,
$$dT = -\frac{4h}{\rho u_m c_p D}(T - T_0) \, dx,$$

and therefore
$$\int_{T_1}^{T} \frac{dT}{(T - T_0)} = -\frac{4hx}{\rho u_m c_p D},$$

i.e.
$$\log_e \frac{T - T_0}{T_1 - T_0} = -\frac{4hx}{\rho u_m c_p D}$$

or
$$(T - T_0) = (T_1 - T_0) \, e^{-\beta x}, \qquad (9.2.1)$$

where
$$\beta = \frac{4h}{\rho u_m c_p D} \quad \text{or} \quad \frac{h\pi D}{m c_p}$$

and
$$m = \text{mass flow} = \rho u_m \frac{\pi D^2}{4}.$$

Equation (9.2.1) gives the temperature distribution with distance along the pipe. Note that we are taking a mean value for the temperature *across* any section of the tube.

The exit temperature at $x = L$ is given by

$$(T_2 - T_0) = (T_1 - T_0) \exp\left(- \frac{h\pi DL}{mc_p}\right). \qquad (9.2.2)$$

The total heat flow is given by $Q = mc_p(T_1 - T_2)$, but from (9.2.2)

$$\log_e \frac{T_2 - T_0}{T_1 - T_0} = -\frac{h\pi DL}{mc_p},$$

therefore

$$mc_p = \frac{h\pi DL}{\log_e \dfrac{T_1 - T_0}{T_2 - T_0}},$$

and the total rate of heat transfer is

$$Q = h\pi DL \frac{(T_1 - T_2)}{\log_e \dfrac{T_1 - T_0}{T_2 - T_0}} \qquad (9.2.3)$$

or

$$Q = h\pi DL \,\Delta T_{Lm}, \qquad (9.2.4)$$

where ΔT_{Lm} is the logarithmic mean-temperature difference defined by

$$\Delta T_{Lm} = \frac{T_1 - T_2}{\log_e \dfrac{T_1 - T_0}{T_2 - T_0}} = \frac{\Delta T_1 - \Delta T_2}{\log_e \dfrac{\Delta T_1}{\Delta T_2}}. \qquad (9.2.5)$$

9.3. Mean temperature difference for a counter-flow heat exchanger

Consider next a counter-flow heat exchanger as illustrated in fig. 65. Let fluid (*a*) be cooled from temperature T_{a1} to T_{a2} and let fluid (*b*) be heated from T_{b1} to T_{b2}.

Overall heat balance:

Writing $W_a = (mc_p)_a$ and $W_b = (mc_p)_b$ for the product of mass flow and specific heat in each case, we can express the overall balance

$$Q = W_a(T_{a1} - T_{a2}) = W_b(T_{b2} - T_{b1}). \qquad (9.3.1)$$

Local heat balance:

Heat given up by stream (*a*) in length $dx = -W_a\, dT_a$ (note the negative sign, since T_a is decreasing as x increases). Heat taken up by stream (b) in length $dx = -W_b\,dT_b$ (note the negative sign, since, although T_b is increasing in the direction of flow of fluid (b), it is decreasing as x increases from left to right in the figure).

Therefore for the local balance,

$$-W_a\,dT_a = -W_b\,dT_b = q\,dA = q\pi D\,dx,$$

i.e.

$$dT_a = -\frac{q\,dA}{W_a} \quad \text{and} \quad dT_b = -\frac{q\,dA}{W_b},$$

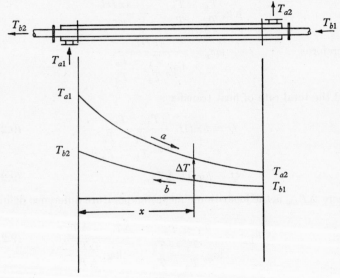

Fig. 65

therefore

$$d(T_a - T_b) = d(\Delta T) = -\left(\frac{1}{W_a} - \frac{1}{W_b}\right)q\,dA$$

$$= -\left(\frac{1}{W_a} - \frac{1}{W_b}\right)q\pi D\,dx,$$

also $q = h_0\Delta T$, where h_0 is the overall heat-transfer factor, therefore

$$\frac{d(\Delta T)}{\Delta T} = -h_0\pi D\left(\frac{1}{W_a} - \frac{1}{W_b}\right)dx, \tag{9.3.2}$$

and integrating from $x = 0$ to $x = L$

$$\log_e \frac{(T_{a2} - T_{b1})}{(T_{a1} - T_{b2})} = -h_0\pi DL\left(\frac{1}{W_a} - \frac{1}{W_b}\right) \tag{9.3.3}$$

or
$$\frac{T_{a2} - T_{b1}}{T_{a1} - T_{b2}} = e^{-\alpha}, \qquad (9.3.4)$$

where
$$\alpha = h_0 \pi D L \left(\frac{1}{W_a} - \frac{1}{W_b} \right).$$

From equation (9.3.1)
$$T_{a1} - T_{a2} = \frac{Q}{W_a},$$

and
$$T_{b2} - T_{b1} = \frac{Q}{W_b},$$

hence,
$$(T_{a1} - T_{b2}) - (T_{a2} - T_{b1}) = Q \left(\frac{1}{W_a} - \frac{1}{W_b} \right),$$

i.e.
$$Q = \frac{(T_{a1} - T_{b2}) - (T_{a2} - T_{b1})}{\left(\dfrac{1}{W_a} - \dfrac{1}{W_b} \right)};$$

substituting for $(1/W_a - 1/W_b)$ from (9.3.3)

$$Q = h_0 \pi D L \frac{(T_{a1} - T_{b2}) - (T_{a2} - T_{b1})}{\log_e \dfrac{(T_{a1} - T_{b2})}{(T_{a2} - T_{b1})}} \qquad (9.3.5)$$

or
$$Q = h_0 \pi D L \, \Delta T_{Lm}, \qquad (9.3.6)$$

i.e. we have the logarithmic mean-temperature difference again.

9.4. Efficiency of a counter-flow heat exchanger

Referring again to fig. 65, if we know the two inlet temperatures T_{a1} and T_{b1} we may wish to calculate the two outlet temperatures T_{a2} and T_{b2} for given values of the mass flows, surface area and heat-transfer factor.

From (9.3.4) $\quad (T_{a2} - T_{b1}) = (T_{a1} - T_{b2}) e^{-\alpha}$,

i.e. $\qquad (T_{a2} - T_{a1}) = (T_{b1} - T_{a1}) + (T_{a1} - T_{b2}) e^{-\alpha}$.

Substituting from (9.3.1)
$$T_{b2} = T_{b1} + \frac{W_a}{W_b} (T_{a1} - T_{a2}),$$

therefore
$$(T_{a2} - T_{a1}) = (T_{b1} - T_{a1}) + (T_{a1} - T_{b1}) e^{-\alpha} - \frac{W_a}{W_b} (T_{a1} - T_{a2}) e^{-\alpha},$$

i.e. $(T_{a1} - T_{a2}) \left\{ 1 - \dfrac{W_a}{W_b} e^{-\alpha} \right\} = (T_{a1} - T_{b1})(1 - e^{-\alpha})$

or $(T_{a1} - T_{a2}) = \eta(T_{a1} - T_{b1}),$ (9.4.1)

where $\eta = \dfrac{1 - e^{-\alpha}}{1 - \dfrac{W_a}{W_b} e^{-\alpha}}.$

Equation (9.4.1) enables the outlet temperature T_{a2} to be calculated if the inlet temperatures T_{b1} and T_{a1} are known. The ratio η is the efficiency of the heat exchanger in cooling fluid (a). Similarly for the outlet temperature T_{b2}

$$(T_{b2} - T_{b1}) = \frac{W_a}{W_b}(T_{a1} - T_{a2}) = \frac{W_a}{W_b} \eta(T_{a1} - T_{b1}),   (9.4.2)$$

and the ratio $W_a\eta/W_b$ is the efficiency of the heat exchanger in heating fluid (b).

There are two cases to consider:

(i) If the heat capacity of the fluid stream (b) is greater than that of (a) ΔT can approach zero at the cold end. This is the case shown on fig. 65, and since $W_b > W_a$, α is positive and $\eta \to 1$ as the surface area $\pi DL \to \infty$, i.e.

maximum value of the ratio $\dfrac{T_{a1} - T_{a2}}{T_{a1} - T_{b1}} = 1,$

maximum value of the ratio $\dfrac{T_{b2} - T_{b1}}{T_{a1} - T_{b1}} = \dfrac{W_a}{W_b}.$

(ii) If the heat capacity of the fluid stream (b) is less than (a) $W_b < W_a$, α is negative and $\eta \to W_b/W_a$ as the surface area $\to \infty$, i.e.

maximum value of the ratio $\dfrac{T_{a1} - T_{a2}}{T_{a1} - T_{b1}} = \dfrac{W_b}{W_a},$

maximum value of the ratio $\dfrac{T_{b2} - T_{b1}}{T_{a1} - T_{b1}} = 1.$

9.5. Parallel-flow heat exchangers

This arrangement is shown in fig. 66. The analysis is similar to that of §9.3. For the *overall heat balance*

$$Q = W_a(T_{a1} - T_{a2}) = W_b(T_{b2} - T_{b1}),   (9.5.1)$$

and for the *local heat balance*

heat given up by stream (a) in length $dx = -W_a\,dT_a,$

heat taken up by stream (b) in length $dx = +W_b\,dT_b,$

hence in place of equation (9.3.2) we have

$$\frac{\mathrm{d}(\Delta T)}{\Delta T} = - h_0 \pi D \left(\frac{1}{W_a} + \frac{1}{W_b} \right) \mathrm{d}x, \tag{9.5.2}$$

hence, integrating from $x = 0$ to $x = L$

$$\log_e \frac{(T_{a2} - T_{b2})}{(T_{a1} - T_{b1})} = - h_0 \pi D L \left(\frac{1}{W_a} + \frac{1}{W_b} \right), \tag{9.5.3}$$

or $\quad \dfrac{\Delta T_2}{\Delta T_1} = \dfrac{T_{a2} - T_{b2}}{T_{a1} - T_{b1}} = \exp \left\{ - h_0 \pi D L \left(\dfrac{1}{W_a} + \dfrac{1}{W_b} \right) \right\}. \quad$ (9.5.4)

For the total rate of heat transfer we again have the result

$$Q = h_0 \pi D L \, \Delta T_{Lm}. \tag{9.5.5}$$

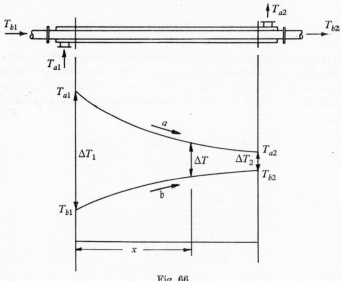

Fig. 66

9.6. Double-pass, multi-pass and cross-flow heat exchangers

Heat exchangers of the shell and tube type are frequently designed with a double- or multi-pass system for flow through the tubes. A double-pass heat exchanger with a hair-pin tube arrangement is shown diagrammatically in fig. 67. The normal procedure is to start with the logarithmic mean-temperature difference defined by

$$\Delta T_{Lm} = \frac{(T_{a1} - T_{b2}) - (T_{a2} - T_{b1})}{\log_e \dfrac{(T_{a1} - T_{b2})}{(T_{a2} - T_{b1})}}, \tag{9.6.1}$$

and then to apply a correction factor, so that the working mean-temperature difference is given by

$$\Delta T_m = Y \Delta T_{Lm}.$$ (9.6.2)

In [1] and [2] the correction factor Y is plotted against the temperature ratio $(T_{b2} - T_{b1})/(T_{a1} - T_{b1})$ for a range of values of the heat-capacity ratio W_b/W_a and for various types of heat exchanger.

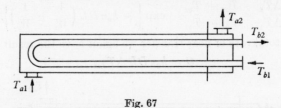

Fig. 67

A cross-flow heat exchanger is represented diagrammatically in fig. 68. For this arrangement we can adopt the same procedure, starting with the logarithmic mean-temperature difference for pure counter-flow defined by (9.6.1) and applying a correction factor.

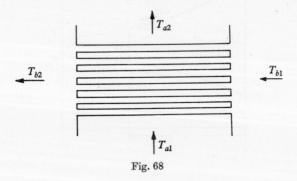

Fig. 68

The reader should again consult [1] or [2] for charts giving values of the correction factor.

If the temperature change of each fluid stream is not too large, however, it is sufficiently accurate for most purposes to take the arithmetic mean-temperature difference for a pure cross-flow heat exchanger, i.e.

$$\Delta T_{am} = \frac{(T_{a1} + T_{a2}) - (T_{b1} + T_{b2})}{2}.$$ (9.6.3)

Many practical cases involve a combination of counter-flow and cross-flow; for instance, shell and tube heat exchangers with baffles inside the shell to direct the flow backwards and forwards across the tube bundle.

REFERENCES

(1)   McAdams. *Heat Transmission* (McGraw-Hill).
(2)   Fishenden and Saunders. *Heat Transfer* (Oxford).

CHAPTER 10

DIMENSIONAL ANALYSIS APPLIED TO HEAT TRANSFER

10.1. Note on dimensions for temperature and heat

The methods of dimensional analysis and the use of the Buckingham Π theorem were illustrated in Chapter 5 in their application to a number of problems in fluid mechanics. We can now apply the same technique to certain problems in heat transfer.

According to the kinetic theory of gases the absolute temperature in a gas is measured by the average kinetic energy of its molecules. In fluid mechanics, however, we are not normally concerned with the motion of individual molecules. We shall therefore introduce a fourth fundamental dimension Θ for *temperature*,† in addition to the three fundamental dimensions M, L and T, which are required for mechanical quantities:

Derived quantity	Dimensions in terms of	
	M, L, T, Θ	M, L, T, Θ, H
Heat quantity	ML^2/T^2	H
Heat-transfer rate, Q	ML^2/T^3	H/T
Heat current density, q	M/T^3	H/L^2T
Heat-transfer factor, h	$M/T^3\Theta$	$H/L^2T\Theta$
Mechanical equivalent of heat, J	—	ML^2/T^2H
Specific heat, c_p	$L^2/T^2\Theta$	$H/M\Theta$
Thermal conductivity, k	$ML/T^3\Theta$	$H/LT\Theta$
Thermal diffusivity, $k/\rho c_p$	L^2/T	L^2/T

According to the first law of thermodynamics, heat and work are interchangeable within certain limits. If the mechanical equivalent of heat J is dimensionless, therefore, *heat* should have dimensions ML^2T^{-2}. In some problems of fluid mechanics and heat transfer there is a measurable interchange of energy between thermal and mechanical quantities, for instance, in the flow of steam through a nozzle. In problems of this sort, therefore, it is appropriate to work with the four fundamental dimensions M, L, T, Θ. In other problems, however, for instance, the flow of a liquid through a heat

† In this chapter the letter θ will be used for temperature to avoid confusion with the dimension of time T. Elsewhere in the book the letter T is used in preference to θ to denote temperature.

exchanger, there is no measurable interchange between thermal and mechanical quantities. It is true that fluid friction will always produce some thermal effect, but in many cases this is insignificant compared with the heat that is being transferred between the walls of the exchanger and the fluid. In problems of this kind, therefore, it will be appropriate to introduce a fifth fundamental dimension H for *heat*.

The table on the facing page lists the dimensions of the additional derived quantities which we shall require for analysing heat transfer on the M, L, T, Θ system and on the M, L, T, Θ, H system.

10.2. Forced convection—flow through a tube with heat transfer

It will be reasonable to assume that the local heat flux q will depend on the diameter of the pipe D, the mean temperature difference $\Delta\theta$ between tube wall and fluid, the mean velocity u_m of the fluid, and on the physical properties k, c_p, ρ and μ, i.e. we will assume that

$$q = f(\Delta\theta, k, c_p, \rho, \mu, D, u_m).$$

By Buckingham's Π theorem, this connexion between eight quantities (one dependent and seven independent variables) should reduce to a relationship between three dimensionless ratios on the M, L, T, Θ, H system.

Using the method of Chapter 5, take primary quantities as follows:

Quantity	$\Delta\theta$	k	ρ	D	u_m
Dimensions	Θ	$H/LT\Theta$	M/L^3	L	L/T

We now require dimensionless ratios for q, c_p and μ. For Π_1 try

$$\frac{q}{\Delta\theta^{a1}k^{b1}\rho^{c1}D^{d1}u_m^{e1}};$$

dimensionally we must have

$$\left[\frac{H}{L^2T}\right] = \left[\Theta\right]^{a1}\left[\frac{H}{LT\Theta}\right]^{b1}\left[\frac{M}{L^3}\right]^{c1}\left[L\right]^{d1}\left[\frac{L}{T}\right]^{e1},$$

hence, for $[M]$ $c_1 = 0$,

$\qquad\qquad [H]$ $b_1 = 1$,

$\qquad\qquad [\Theta]$ $a_1 = b_1 = 1$,

$\qquad\qquad [T]$ $b_1 + e_1 = 1$, $\therefore e_1 = 0$,

$\qquad\qquad [L]$ $b_1 + 3c_1 - d_1 - e_1 = 2$, $\therefore d_1 = -1$,

therefore

$$\Pi_1 = \frac{qD}{\Delta\theta k} \quad\text{or}\quad \frac{hD}{k}, \quad \text{the *Nusselt number*, } Nu.$$

By a similar procedure we arrive at

$$\Pi_2 = \frac{c_p \rho u_m D}{k}, \quad \text{which is the } \textit{Péclet number, Pe,}$$

and $\Pi_3 = \dfrac{\mu}{\rho u_m D}$,   the reciprocal of the *Reynolds number*.

We can therefore say that for forced convection in a tube the Nusselt number is a function of the Péclet and Reynolds numbers, i.e.

$$Nu = f(Pe, Re). \tag{10.2.1}$$

The Péclet number $\dfrac{c_p \rho u_m D}{k} = \dfrac{\mu c_p}{k} Re.$

$\mu c_p / k$ is another dimensionless ratio the *Prandtl number, Pr*. We can therefore say, as an alternative to (10.2.1), that the Nusselt number is a function of the Reynolds and Prandtl numbers, i.e.

$$Nu = f(Re, Pr). \tag{10.2.2}$$

It is preferable to use the combination of Reynolds and Prandtl numbers as in (10.2.2) because, unlike the Péclet number, the Prandtl number involves only the physical properties of the fluid and not u_m or D.

If we had chosen a different selection of primary quantities, for instance, $\Delta\theta$, c_p, ρ, D, u_m, we would have found that

$$\Pi_1 = \frac{q}{\Delta\theta \rho u_m c_p} \quad \text{or} \quad \frac{h}{\rho u_m c_p}, \quad \text{the } \textit{Stanton number, St,}$$

and in place of (10.2.1) or (10.2.2) we would have

$$St = f(Re, Pr). \tag{10.2.3}$$

The Nusselt and Stanton numbers are alternative dimensionless heat-transfer coefficients for forced convection. Note that $St = Nu/RePr$.

It is found experimentally that for the case of fully developed *turbulent flow* in a long tube, with $Re \gg 2100$, and for fluids whose Prandtl numbers are in the range from about 0·5 upwards, heat-transfer rates are given within an accuracy of about ± 10 per cent by the simple empirical expression

$$Nu = 0{\cdot}023 Re^{0{\cdot}8} Pr^{0{\cdot}4}. \tag{10.2.4}$$

The heat-transfer factor h in the Nusselt number will be a mean value and, as shown in the preceding chapter, is given by the average heat flux divided by the logarithmic mean-temperature difference between tube wall and fluid. The physical properties μ, c_p and k are usually evaluated at the mean bulk temperature of the fluid, i.e. at the arithmetic mean of inlet and outlet temperatures. Minor variations on equation (10.2.4) have been suggested, however, which claim to give better accuracy and involve the evaluation of the physical properties of the fluid at some hypothetical 'mean film temperature'. Equation (10.2.4) is of surprisingly wide application and covers most ordinary gases and liquids encountered in chemical engineering. It can be applied to heat-transfer calculations with passages or channels whose cross-sections are other than circular, provided appropriate values are taken for the equivalent diameter. However, the accuracy of the calculation is generally less satisfactory in such cases.

For fully developed *laminar flow* in a circular pipe with uniform wall temperature, an analytical solution is possible if we assume constant viscosity and a parabolic velocity distribution. The result may be expressed approximately by

$$Nu = \frac{h_{av}D}{k} = 1 \cdot 62 \left(RePr \frac{D}{L} \right)^{\frac{1}{3}}. \tag{10.2.5}$$

In this case we take the arithmetic mean-temperature difference in evaluating the Nusselt number. In practice the viscosity of a liquid varies appreciably with temperature and the velocity distribution will not be parabolic. The following empirical modification of (10.2.5) is found to give better agreement with experimental results

$$Nu = 1 \cdot 86 \left(\frac{\mu}{\mu_s} \right)^{0 \cdot 14} \left(RePr \frac{D}{L} \right)^{\frac{1}{3}}. \tag{10.2.6}$$

μ_s is the viscosity at the wall temperature, and μ is the value at the mean bulk temperature of the fluid.

10.3. Physical significance of the Nusselt, Stanton and Prandtl numbers

It is important to appreciate the physical significance of the dimensionless ratios employed in heat-transfer calculations.

The *Nusselt number* is a dimensionless heat-transfer coefficient which gives a measure of the ratio of the heat-transfer rate q to the rate at which heat would be conducted within the fluid under a temperature gradient $\Delta\theta/D$.

The *Stanton number* is an alternative heat-transfer coefficient and gives a measure of the ratio of the heat-transfer factor h or $q/\Delta\theta$ to the flow of heat along the pipe per unit temperature rise, due to the velocity and heat capacity of the fluid, i.e. to $\rho u_m c_p$.

The *Prandtl number* $\mu c_p/k = \nu/\alpha$ and is therefore simply the ratio of kinematic viscosity to thermal diffusivity.

The significance of the *Reynolds number* as a measure of the ratio of inertia forces to viscous forces in the flow has already been mentioned in Chapter 5. An alternative derivation of these dimensionless ratios and their relationship to the necessary conditions for similarity will be given in Chapters 12 and 16.

In all problems of forced convection, provided high velocities of flow are not involved, and provided the physical properties of the fluid, such as density, viscosity and thermal conductivity, are substantially constant, we can say that heat-transfer rates expressed in non-dimensional form as a Nusselt or Stanton number are functions only of the Reynolds number and the Prandtl number. The form of the function will of course depend on the geometry of the system and, although in a few special cases of laminar flow it may be calculated, in most cases it must be determined by experiment. Equations (10.2.4) and (10.2.5) are particular examples of this functional relationship for the case of a long tube of circular cross-section. Appropriate working formulae for other geometrical arrangements, such as flow across single tubes and tube banks, are given by McAdams in [1]. Some of these formulae will be discussed in connexion with numerical examples.

If we are dealing with problems of heat transfer in which there are large temperature differences it may be necessary to take account of the variation in value of the physical properties of the fluid with temperature. We should then include a characteristic reference temperature, for instance, the mean bulk temperature of the fluid, θ_m, in addition to the temperature difference $\Delta\theta$ in the dimensional analysis of §10.2. This involves the introduction of another dimensionless ratio $\Delta\theta/\theta_m$, so that in place of the basic equation (10.2.2) we now have

$$Nu = f(Re, Pr, \Delta\theta/\theta_m). \qquad (10.3.1)$$

Before proceeding further with other types of heat-transfer mechanism, it will be appropriate to give a few numerical examples of heat-transfer calculations for forced convection.

Example 1. A surface condenser is to be designed to handle 50,000 lb./hr. of steam at 1 p.s.i. absolute pressure. The condenser is to be of the normal shell and tube type with $\frac{3}{4}$ in. outer diameter by 18 s.w.g. brass tubes (0·654 in. inner diameter) and length 10 ft. between tube plates. Under the worst conditions cooling water is available at 70°F. and the outlet temperature is to be 90°F. Water velocity inside the tubes is to be approximately 6 ft./sec. For heat transfer take $h = 2000$ B.TH.U./hr. sq.ft. °F. on the steam side,

$Nu = 0.023Re^{0.8}Pr^{0.4}$ on the water side, and allow for a fouling factor of 0.0010 hr. sq.ft. °F./B.TH.U. referred to the water side.

Calculate a suitable number of tubes and the number of passes on the water side.

For steam condensing at 1 p.s.i. abs. the saturation temperature is $101.7°$F. and the latent heat is 1036 B.TH.U./lb.

The total heat load is therefore

$$50,000 \times 1036 = 51.8 \times 10^6 \text{ B.TH.U./hr.}$$

The logarithmic mean-temperature difference $= \dfrac{20}{\log_e \dfrac{31.7}{11.7}}$

$$= 20 \text{ approx.}$$

The mass flow of water required $= \dfrac{51.8 \times 10^6}{3600 \times 20} = 718$ lb./sec.

Volume flow of water $= 11.5$ cusecs.

For water at a mean temperature of $80°$F.,

$$\nu = 0.936 \times 10^{-5} \text{ sq.ft./sec.}$$
$$k = 0.35 \text{ B.TH.U./ft.hr. °F.}$$
$$Pr = 5.93$$

and $$Re = \dfrac{6 \times 0.0545}{0.936 \times 10^{-5}} \quad \text{or} \quad Re = 35,000.$$

For heat transfer on the water side, therefore,

$$Nu = 203 \quad \text{and} \quad h_w = 1300 \text{ B.TH.U./hr. sq.ft. °F.}$$

The heat-transfer factor for the condensing steam is 2000 referred to the outer diameter of the tubes, or 2295 referred to the inside diameter.

The overall heat-transfer factor, referred to the inside surface of the tubes, will therefore be given by

$$\frac{1}{h_{\text{overall}}} = \tfrac{1}{1300} + 0.0010 + \tfrac{1}{2295} = 0.00221,$$

therefore $\quad h_{\text{overall}} = 453$ B.TH.U./hr. sq.ft. °F.

Total surface area required $= \dfrac{51.8 \times 10^6}{20 \times 453} = 5720$ sq.ft.

Inner surface area of one tube of 10 ft. length $= 1.712$ sq.ft., therefore

$$\text{number of tubes required} = 3340.$$

Number of tubes in one pass for a water flow of 11·5 cusecs and a velocity of approximately 6 ft./sec. = 822.

A *four-pass arrangement* would therefore be suitable on the water side.

It will be noted in the above example that the fouling factor represents the biggest individual resistance to heat transfer. This is frequently the case with heat exchangers used in chemical plants. It cannot be too strongly emphasized that heat-transfer formulae such as equations (10.2.4) and (10.2.6) are based on measurements with *clean surfaces*. Industrial heat-transfer equipment inevitably undergoes deterioration in service with the progressive accumulation of dirt on the surfaces. How rapidly this deterioration occurs will depend entirely on the nature of the fluids which are being handled. Arrangements must always be made for the periodic cleaning of heat exchangers. Appropriate values for the fouling factors to be used in design calculations must be based on previous operating experience with similar plant.

Example 2. An economizer is to be designed for a boiler which evaporates 100,000 lb. of water/hr. The feed water is to be heated from 150 to 250° F. by means of 150,000 lb./hr. of furnace gases which enter the economizer section at a temperature of 700° F.

The economizer is to consist of a bank of staggered tubes fitting in a duct having a cross-section 10 by 20 ft. with the furnace gas flowing across the tubes. The tubes are of steel and have a length of 20 ft., inside diameter 2·0 in., and outside diameter 2·4 in. There is a clear space of 2·4 in. between adjacent tubes in a row. The tubes in each row are connected to headers so that the water flows through them in parallel, but successive rows are arranged in series. Estimate the number of tube rows required.

For the furnace gases assume thermal properties of air at the appropriate mean temperature. For heat transfer on the gas side, with the above tube spacing, take $Nu = 0·34Re^{0·6}$. For heat transfer on the water side assume $Nu = 0·023Re^{0·8}Pr^{0·4}$. Thermal conductivity of steel = 26 B.TH.U./ft.hr. °F.

The total heat load in this case is approximately 10^7 B.TH.U./hr.

The feed water enters at 150° F., leaves at 250° F., and has a mean-temperature of 200°F.

The furnace gases enter at 700° F., leave at 433° F., and have a mean temperature of approximately 566° F.

Referring to Chapter 9, §9.4, we can take the arithmetic mean temperature difference for the heat-transfer calculation in this case without serious error, i.e. $\Delta\theta = 366°$ F.

For the gas flow. With a spacing of one diameter between adjacent tubes, there will be 24 or 25 tubes in one tube-row of 10 ft. width.

The minimum cross-sectional area available for the gas flow will be approximately $20 \times 5 = 100$ sq.ft. The mass flow of gas is 41·6 lb./sec., and with a density of 0·039 lb./cu.ft. at the mean temperature, the volume flow will be 1068 cu.ft./sec., i.e. the velocity between tubes = 10·7 ft./sec.

In evaluating the Reynolds number for the gas flow across a tube bank it is customary to take the velocity through the narrowest cross-section between tubes and to take the tube diameter as the characteristic length. The kinematic viscosity v should be taken at the mean film temperature. Guessing that the gas side resistance predominates and that the mean tube-wall temperature will be fairly close to 200°F., therefore, we will assume a mean film temperature of 383 °F. At this temperature the value of v is $3·68 \times 10^{-4}$. The Reynolds number for the gas flow will therefore be given by

$$Re = \frac{10·7 \times 0·2}{3·68 \times 10^{-4}} = 5800,$$

and
$$Nu = 0·34 Re^{0·6} = 61·5.$$

[Note that this expression for the Nusselt number can only be used for calculations with air or gases of similar Prandtl number. The effect of Prandtl number is included in the numerical constant 0·34. The formula we are using is still a special case of the general equation (10.2.2), i.e. Nu is a function of Re and Pr. Strictly speaking, since the temperature difference is relatively large and since we have to evaluate the physical properties of the gas at a hypothetical mean film temperature, the basic equation in this case should be (10.3.1), i.e.

$$Nu = f\left(Re,\ Pr,\ \frac{\Delta\theta}{\theta_m}\right).$$

If we were attempting a more accurate calculation or a closer correlation of different experimental results we would have to evolve a more general formula for the heat-transfer coefficient bringing in the temperature ratio term in addition to Re and Pr. In most cases, however, heat-transfer calculations are not required to a very high order of accuracy, and the procedure used in this example is typical of the kind of approximation usually employed. Reference should be made to McAdams or to Fishenden and Saunders [1, 2] for the appropriate working formulae for heat transfer with tube banks having different geometrical arrangements.]

Taking the thermal conductivity of the gas at the mean film temperature, the heat-transfer factor on the gas side will be given by

$$h_g = 6·09 \text{ B.TH.U./hr. sq.ft. °F.}$$

At this point we can make a quick assessment of the surface area required. Assuming that the resistance on the gas side predominates, and guessing an overall $h = 6.0$, area required $= \dfrac{10^7}{6 \times 366}$ $= 4550$ sq.ft. The value of $h = 6$ is typical for gas flow at moderate velocities across tube banks.

For the water flow. Taking parallel flow through 25 tubes in a tube-row, the mass flow through each tube will be 4000 lb./hr. or 1·11 lb./sec. The velocity in the tubes is then given by 0·825 ft./sec. With ν evaluated at 200° F., the Reynolds number for flow inside the tubes is given by

$$Re = 4.02 \times 10^4.$$

Hence, the Nusselt number

$$Nu = 0.023 Re^{0.8} Pr^{0.4} = 143$$

and the heat-transfer factor

$$h_w = 335 \text{ B.TH.U./hr. sq.ft. °F.}$$

This again is a typical value for the heat-transfer factor for water flowing inside pipes with velocities of the order of 1 ft./sec.

If we now include the effect of conduction through the tube wall using the method of Chapter 8, §8.5, we arrive at a value for the overall heat-transfer coefficient (referred to the outer surface area of the tubes) given by

$$h_{\text{overall}} = 5.94 \text{ B.TH.U./hr. sq.ft. °F.}$$

The surface area required is therefore 4600 sq.ft.

Say 15 tube-rows giving 4700 sq.ft.

Suppose that, owing to the use of untreated feed water, a scale deposit of thickness 0·1 in. forms on the inside of the tubes. Calculate the new outlet temperature of the feed water leaving the economizer if k for the scale deposit $= 0.25$ B.TH.U./hr.ft. °F.

By the same method of calculation as before, the new value for the overall heat transfer factor $h = 4.75$ B.TH.U./hr. sq.ft. °F. For a first approximation, assume the same mean-temperature difference as before. The total rate of heat transfer would then be 8.17×10^6 B.TH.U./hr. The temperature rise of the feed water would then be about 82° F. and the new delivery temperature 232° F. Taking this last value, we can work out a new mean-temperature difference and carry out a second approximation.

Example 3. Liquid sodium is to be used as a heat-transfer fluid in a heat-exchanger system. Taking a sodium velocity of 10 ft./sec. and a mean temperature of 400°C., and assuming flow through tubes having an equivalent diameter of $\frac{1}{2}$ in., estimate the heat-transfer factor h.

The following properties may be assumed for sodium at 400°C.: $\rho = 53\cdot4$ lb./cu.ft., $c_p = 0\cdot305$, $\mu = 2\cdot74 \times 10^{-4}$ lb./ft.sec., and $k = 62$ B.TH.U./ft.hr. °F. The Reynolds and Prandtl numbers are then evaluated as follows:

$$Re = 8\cdot12 \times 10^4, \quad Pr = 0\cdot00485.$$

The numerical value of the Reynolds number indicates normal turbulent flow conditions. The value of the Prandtl number, however, is very low compared with typical values for conventional fluids such as air and water. A low value for the Prandtl number is a feature of any liquid metal and it simply reflects the relative importance of the thermal conductivity. The ordinary expression (10.2.4) for heat transfer in turbulent flow cannot be used under these conditions because it is not appropriate to such a different range of Prandtl number. The following formula, due to Lyon, may be employed instead:

$$Nu = 7 + 0\cdot025Re^{0\cdot8}Pr^{0\cdot8}. \tag{10.3.2}$$

Hence, with numerical values for the example considered, $Nu = 9\cdot98$ and the heat-transfer factor $h = 14{,}800$ B.TH.U./ft.²hr. °F.

10.4. Forced convection in high-speed flow

If thermal energy and directed kinetic energy are partly interchangeable, for instance, in the case of the flow of a gas through a nozzle or diffuser, we can no longer attribute independent dimensions to heat. It will therefore be of interest to apply the same method of dimensional analysis for forced convection in a tube but using only the four fundamental dimensions M, L, T, Θ.

We must now assume that

$$q = f(\Delta\theta, k, c_p, \rho, \mu, D, u_m, \theta_m).$$

If we take primary quantities as follows, noting that we are now limited to four primary quantities,

quantity	$\Delta\theta$	k	D	u_m
dimensions	Θ	$ML/T^3\Theta$	L	L/T

we can find dimensionless ratios for the remaining quantities by the usual procedure. The result is as follows:

for q $\Pi_1 = \dfrac{qD}{\Delta\theta k}$,   which is the *Nusselt number* again;

for c_p $\Pi_2 = \dfrac{c_p\Delta\theta}{u_m^2}$ or $\dfrac{Jc_p\Delta\theta}{u_m^2}$,   since J is now dimensionless;

for ρ $\Pi_3 = \dfrac{\rho u_m^3 D}{\Delta\theta k}$;

for μ $\Pi_4 = \dfrac{\mu u_m^2}{\Delta\theta k}$;

for θ_m $\Pi_5 = \dfrac{\theta_m}{\Delta\theta}$,   or the reciprocal $\dfrac{\Delta\theta}{\theta_m}$.

Noting that $\Pi_2\Pi_3 = JRePr$ and that $\Pi_2\Pi_4 = Pr$, it is clear that the only new number involved is the ratio Π_2 or its reciprocal. For convenience we will choose the reciprocal $u_m^2/Jc_p\Delta\theta$. We can therefore say that, in place of equation (10.3.1), for heat transfer by forced convection in high-speed flow

$$Nu = f\left(Re,\ Pr,\ \frac{u_m^2}{Jc_p\Delta\theta},\ \frac{\Delta\theta}{\theta_m}\right). \tag{10.4.1}$$

The physical significance of the new ratio $u_m^2/Jc_p\Delta\theta$ will be clear from the fact that $\tfrac{1}{2}u_m^2$ is the kinetic energy per unit mass of the stream, while $Jc_p\Delta\theta$ is a measure of the enthalpy change per unit mass corresponding to a rise in temperature of the stream by an amount equal to the imposed temperature difference $\Delta\theta$.

The significance of the new ratio will also be apparent when it is recalled that the equation for the one-dimensional compressible flow of a perfect gas, in the absence of heat transfer and shear work, is

$$Jc_p\theta + \tfrac{1}{2}u^2 = \text{constant.}$$

We can therefore say that $u_m^2/2Jc_p$ gives the order of magnitude of the temperature variations that can occur in the flow as a result of compressibility effects. At low speeds this dynamic temperature difference is usually small compared with the imposed temperature difference $\Delta\theta$. In high-speed flow, however, it may be of the same order of magnitude.

For a perfect gas we can write

$$Jc_p = \frac{\gamma}{\gamma-1}\,R,$$

therefore $Jc_p\Delta\theta = \dfrac{\gamma}{\gamma-1}\,R\theta_m\,\dfrac{\Delta\theta}{\theta_m} = \dfrac{a_m^2}{\gamma-1}\,\dfrac{\Delta\theta}{\theta_m}$, where $a_m = \sqrt{(\gamma R\theta_m)}$ = speed of sound in the gas at temperature θ_m. The ratio $\dfrac{u_m^2}{Jc_p\Delta\theta}$ is therefore equal to $(\gamma - 1)\dfrac{\theta_m}{\Delta\theta}Ma^2$, where $Ma = u_m/a_m$, the *Mach number* for the flow at the reference point. As an alternative to equation (10.4.1) we can therefore write

$$Nu = f\left(Re,\ Pr,\ Ma,\ \frac{\Delta\theta}{\theta_m}\right). \qquad (10.4.2)$$

10.5. Free convection

In heat transfer by free convection the motion of the fluid is caused by density changes. Let the bulk fluid temperature be θ_0 and the corresponding density ρ_0. The buoyancy force per unit volume for an element of fluid at temperature θ and density ρ will then be $(\rho_0 - \rho)g$, i.e.,

$$\text{buoyancy force per unit mass} = \frac{(\rho_0 - \rho)g}{\rho}.$$

If β is the coefficient of thermal expansion, referred to the bulk temperature of the fluid,

$$\frac{1}{\rho} = \frac{1}{\rho_0}(1 + \beta\Delta\theta)$$

or

$$\rho_0 = \rho(1 + \beta\Delta\theta),$$

i.e. buoyancy force per unit mass $= \beta g(\theta - \theta_0)$ or $\beta g\Delta\theta$. For a perfect gas $\dfrac{\rho_0}{\rho} = \dfrac{p_0\theta}{p\theta_0} \approx \dfrac{\theta}{\theta_0}$ (if we neglect pressure variation), i.e.

$$\text{buoyancy force} = \frac{(\theta - \theta_0)g}{\theta_0} = \frac{\Delta\theta g}{\theta_0}.$$

We will now assume that the heat flux q for the case of heat transfer by *free convection from a vertical plate* will depend on the following variables including the height of the plate L:

$$q = f(\Delta\theta,\ \beta g,\ \rho,\ k,\ c_p,\ \mu,\ L).$$

This connexion between eight quantities should reduce to a relationship between three dimensionless ratios on the M, L, T, Θ, H system. Select primary quantities as follows:

quantity:	$\Delta\theta$	k	ρ	μ	L
dimensions:	Θ	$H/LT\Theta$	M/L^3	M/LT	L

then for q, $\quad \Pi_1 = \dfrac{qL}{\Delta\theta k} = \dfrac{hL}{k}$, the *Nusselt number*, Nu, for βg, $\Pi_2 = \dfrac{\beta g \Delta \theta \rho^2 L^3}{\mu^2}$ or $\dfrac{\beta g \Delta \theta L^3}{\nu^2}$, the *Grashof number, Gr*,

and for c_p, $\Pi_3 = \dfrac{\mu c_p}{k}$, the *Prandtl number, Pr*.

For free convection from a vertical plate, therefore,

$$Nu = f(Gr, Pr). \tag{10.5.1}$$

A theoretical solution, assuming laminar motion in the free convection currents rising from the plate, gives

$$Nu = 0 \cdot 52 Gr^{\frac{1}{4}} Pr^{\frac{1}{4}}.$$

It is found experimentally that, for laminar motion with the product $Gr\,Pr$ in the range from 10^4 to 10^8, the average heat-transfer coefficient is given by

$$Nu = 0 \cdot 56 Gr^{\frac{1}{4}} Pr^{\frac{1}{4}}. \tag{10.5.2}$$

For values of the product $Gr\,Pr$ greater than 10^9 the motion is generally turbulent, and it is found that the Nusselt number is then proportional to $(Gr\,Pr)^{\frac{1}{3}}$. The following working formulae have been suggested by Saunders[2]:

for gases $Nu = 0 \cdot 12 Gr^{\frac{1}{3}} Pr^{\frac{1}{3}},$
for liquids $Nu = 0 \cdot 17 Gr^{\frac{1}{3}} Pr^{\frac{1}{3}}.$ $\Bigg\}$ (10.5.3)

The product of Grashof number and Prandtl number which appears in all these expressions is sometimes known as the *Rayleigh number, Ra*.

For heat transfer by free convection from *horizontal cylinders*, the diameter is used in defining the Nusselt and Grashof numbers, and for values of the product $Gr\,Pr < 10^8$ the following expression may be used:

$$Nu = 0 \cdot 47 Gr^{\frac{1}{4}} Pr^{\frac{1}{4}}. \tag{10.5.4}$$

REFERENCES

(1) MCADAMS. *Heat Transmission* (McGraw-Hill).
(2) FISHENDEN and SAUNDERS. *Heat Transfer* (Oxford).

CHAPTER 11

HEAT TRANSFER AND SKIN FRICTION IN TURBULENT FLOW

11.1. Reynolds analogy

In most practical heat-transfer problems the flow is mainly turbulent. The mechanism of turbulence will be discussed in some detail in Chapter 14, but it will be useful in the meantime to have a simple picture from which certain conclusions may be drawn regarding the relationship between heat transfer and skin friction in turbulent flow.

Consider the case of turbulent flow in a boundary layer or near the wall of a pipe where the mean velocity at any point, averaged over

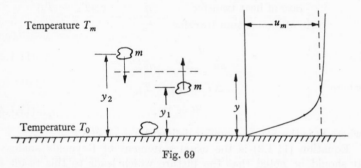

Fig. 69

a short period of time, may be expressed as a function only of the distance y from the wall, i.e.

$$\bar{u} = f(y). \tag{11.1.1}$$

The random turbulent fluctuations in velocity will be superimposed on this mean flow. It is convenient to picture the turbulent motion being produced by the movement of small lumps or particles of fluid backwards and forwards across the stream. Owing to the transverse velocity gradient, this movement must involve the transfer of momentum and hence turbulent shearing stresses are set up in the fluid. Similarly, if there is a transverse temperature gradient, heat transfer must take place by the same mechanism of turbulent convection.

Referring to fig. 69, let $\bar{u}_m$ be the mean or bulk velocity of the flow averaged with respect to distance across the stream. At the wall the

141

velocity must be zero to satisfy the condition of zero slip. Suppose a particle of fluid of mass m moves from a region at distance y_1 from the wall to another region at distance y_2 and that it carries with it the momentum and temperature appropriate to level y_1. At the same time, to satisfy the condition of continuity, we must suppose that another particle of mass m moves from level y_2 to y_1 carrying with it the momentum and temperature appropriate to level y_2. The *net* transfer across the dotted surface shown in fig. 69 will then be as follows:

$$\text{transfer of } x\text{-momentum (inwards)} = m(\bar{u}_2 - \bar{u}_1),$$
$$\text{transfer of heat (outwards)} \quad = - mc_p(T_2 - T_1).$$

If the particles of fluid move on the average from a region where the mean velocity is $\bar{u}_m$ and the bulk temperature is T_m, to the surface where the mean velocity is zero and the bulk temperature is T_0, we can say that

$$\frac{\text{rate of heat transfer}}{\text{rate of momentum transfer}} = \frac{q}{\tau_0} = - \frac{c_p(T_m - T_0)}{\bar{u}_m},$$

i.e.
$$\frac{q}{\rho \bar{u}_m c_p \Delta T} = \frac{\tau_0}{\rho \bar{u}_m^2}, \tag{11.1.2}$$

where
$$\Delta T = T_0 - T_m$$

or
$$St = \tfrac{1}{2}c_f, \tag{11.1.3}$$

where St is the *Stanton number* defined by $q/\rho \bar{u}_m c_p \Delta T$.

Equation (11.1.3) is the usual statement of Reynolds analogy. It should be noted that the picture which leads to this result is over-simplified, not only in the mechanism postulated for turbulent flow but also because no account is taken of the existence of any laminar sub-layer.

11.2. Taylor–Prandtl analogy

In the Taylor-Prandtl analogy the previous method is extended to include the effect of a laminar-flow region adjacent to the wall. The flow is now pictured as being divided into two zones as shown in fig. 70. A laminar sub-layer is assumed to extend from the wall to a thickness δ_L. In this region only viscous shearing stresses are supposed to exist, and heat transfer is maintained by the mechanism of conduction alone. Outside the laminar sub-layer the flow is assumed to be entirely turbulent with negligible viscous stresses. In this outer region shearing stresses arise through the transfer of momentum by the turbulent motion of small particles of fluid, while heat is similarly transferred by the mechanism of turbulent convection alone. In the *laminar sub-layer*, assuming a linear distribution of velocity and temperature,

$$\tau_0 = \mu \frac{\mathrm{d}u}{\mathrm{d}y} = \mu \frac{u_L}{\delta_L},$$

and

$$q = -k \frac{\mathrm{d}T}{\mathrm{d}y} = -\frac{k(T_L - T_0)}{\delta_L},$$

hence,

$$\frac{q}{\tau_0} = -\frac{k(T_L - T_0)}{\mu u_L} = -\frac{c_p(T_L - T_0)}{Pr\, u_L}, \qquad (11.2.1)$$

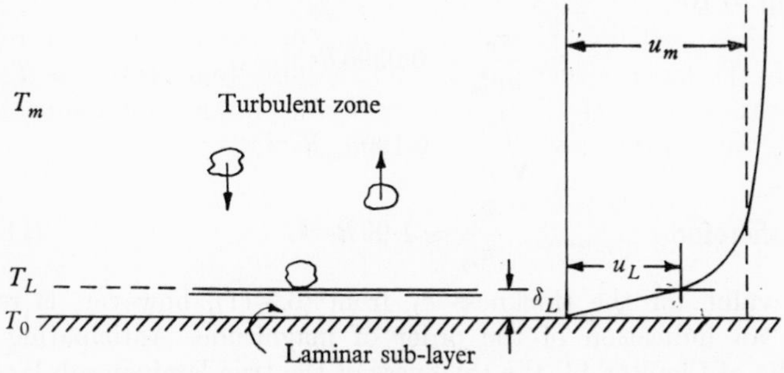

Fig. 70

where Pr is the Prandtl number $\mu c_p / k$. In the *turbulent zone*, assuming that Reynolds analogy can be applied for movement of particles from a region with average properties u_m and T_m to the edge of the laminar sub-layer where the velocity is u_L and the temperature T_L, we can say that

$$\frac{q}{\tau_0} = -\frac{c_p(T_m - T_L)}{(u_m - u_L)}. \qquad (11.2.2)$$

Hence, for transfer at the edge of the laminar sub-layer, eliminating T_L between equations (11.2.1) and (11.2.2),

$$\frac{q}{\tau_0}[(u_m - u_L) + Pr\, u_L] = c_p(T_0 - T_m),$$

i.e.

$$\frac{q}{\rho u_m c_p \Delta T} = \frac{\tau_0}{\rho u_m^2} \frac{1}{1 + \dfrac{u_L}{u_m}(Pr - 1)} \qquad (11.2.3)$$

or

$$St = \frac{\frac{1}{2}c_f}{1 + \dfrac{u_L}{u_m}(Pr - 1)}. \qquad (11.2.4)$$

In the special case of the Prandtl number being numerically equal to 1·0, this reduces to the Reynolds analogy (11.1.3).

An indication of the magnitude of the velocity ratio u_L/u_m may be obtained as follows for the case of flow in a long pipe. From fig. 70 for the laminar sub-layer, $u_L = \dfrac{\tau_0 \delta_L}{\mu}$, and from (6.3.11) an approximate value for δ_L is $\dfrac{10\nu}{\sqrt{(\tau_0/\rho)}}$,

hence,
$$u_L \approx 10 \sqrt{\frac{\tau_0}{\rho}}.$$

But from (6.3.1) for turbulent flow in smooth pipes with values of Re up to 10^5,

$$\frac{\tau_0}{\rho u_m^2} = 0\cdot 0395 Re^{-\frac{1}{4}},$$

i.e.
$$\sqrt{\frac{\tau_0}{\rho}} = 0\cdot 199 u_m Re^{-\frac{1}{8}},$$

and therefore
$$\frac{u_L}{u_m} \approx 1\cdot 99 Re^{-\frac{1}{8}}. \qquad (11.2.5)$$

The value for the thickness δ_L from (6.3.11), however, is really only an indication of the order of magnitude. Anticipating the results of Chapter 14, the thickness of the true laminar sub-layer is probably about half this value, i.e. $\dfrac{5\nu}{\sqrt{(\tau_0/\rho)}}$, and this would correspond to a velocity ratio given by $u_L/u_m \approx 1\cdot 0 Re^{-\frac{1}{8}}$. Complications arise, however, for two reasons. In the first place the sharp division into a laminar sub-layer and an exclusively turbulent zone outside is still an over-simplification, although it is an improvement on the simple Reynolds analogy. Secondly, the velocity and temperature boundary layers are only similar in profile for the special case of the Prandtl number being equal to 1·0. It is therefore necessary to make an empirical adjustment to (11.2.5) to include the effect of Prandtl number variation as well as the smaller value for δ_L. The following expression, due to Hoffmann, may be used:

$$\frac{u_L}{u_m} = 1\cdot 5 Re^{-\frac{1}{8}} Pr^{-\frac{1}{8}}. \qquad (11.2.6)$$

For flow in smooth pipes, with values of Re up to 10^5, therefore, substituting for c_f from (6.3.1) and for the velocity ratio u_L/u_m from (11.2.6), the Taylor-Prandtl equation (11.2.4) becomes

$$St = \frac{0\cdot 0395 Re^{-\frac{1}{4}}}{1 + 1\cdot 5 Re^{-\frac{1}{8}} Pr^{-\frac{1}{8}}(Pr - 1)}, \qquad (11.2.7)$$

and this is found to give good agreement with experimental results for heat transfer, provided the value of the Prandtl number is not greatly different from $1 \cdot 0$ which is the case for most gases.

For air in the range of Reynolds number from 5000 to 50,000, (11.2.7) can be represented approximately by the simpler expression

$$St = 0 \cdot 046 Re^{-\frac{1}{4}} \tag{11.2.8}$$

or, in terms of the Nusselt number

$$Nu = 0 \cdot 046 Re^{\frac{3}{4}} Pr. \tag{11.2.9}$$

11.3. Pressure drop and heat transfer for flow in a tube

Consider the case of a fluid flowing through a tube of diameter D and length L. Let the fluid enter at temperature T_1 and leave at temperature T_2. Provided the temperature change is not too great we can take mean values for the properties of the fluid at the arithmetic mean between T_1 and T_2. Let the average velocity be u_m corresponding to the mean density ρ. Let the tube wall temperature be T_0 and the logarithmic mean-temperature difference be ΔT_{Lm} as defined by (9.2.5). The following statements are quite general and involve only the definitions of c_f and St:

pressure drop:
$$\Delta p = 2c_f \frac{L}{D} \rho u_m^2$$

or
$$\frac{\Delta p}{\rho u_m^2} = 2c_f \frac{L}{D}; \tag{11.3.1}$$

heat transfer:
$$Q = \pi D L q = \rho u_m \frac{\pi D^2}{4} c_p (T_2 - T_1),$$

therefore
$$(T_2 - T_1) = \frac{4q}{\rho u_m c_p} \frac{L}{D}$$

or
$$\frac{T_2 - T_1}{\Delta T_{Lm}} = 4St \frac{L}{D}. \tag{11.3.2}$$

Eliminating L/D between (11.3.1) and (11.3.2)

$$\frac{T_2 - T_1}{\Delta T_{Lm}} = 2 \frac{St}{c_f} \frac{\Delta p}{\rho u_m^2} \tag{11.3.3}$$

or
$$u_m^2 = 2 \frac{St}{c_f} \frac{\Delta p}{\rho} \frac{\Delta T_{Lm}}{T_2 - T_1},$$

i.e.
$$u_m = \sqrt{\left(2 \frac{St}{c_f} \frac{\Delta p}{\rho} \frac{\Delta T_{Lm}}{T_2 - T_1} \right)}. \tag{11.3.4}$$

This is a convenient equation for calculating the maximum permissible velocity that may be used to achieve a given value of $\dfrac{T_2 - T_1}{\Delta T_{Lm}}$, subject to a specified maximum permissible pressure drop Δp. The Reynolds number corresponding to the velocity of the fluid calculated from (11.3.4) is given by

$$Re = \frac{u_m D}{\nu} = \sqrt{\left(2 \frac{St}{c_f} \frac{D^2}{\nu^2} \frac{\Delta p}{\rho} \frac{\Delta T_{Lm}}{T_2 - T_1} \right)}. \qquad (11.3.5)$$

The length of tube required can be calculated from (11.3.2), i.e.

$$\frac{L}{D} = \frac{1}{4St} \frac{T_2 - T_1}{\Delta T_{Lm}},$$

For air and for most permanent gases we can substitute for the Stanton number from (11.2.8) giving

$$\frac{L}{D} = \frac{Re^{\frac{1}{4}}}{0.184} \frac{T_2 - T_1}{\Delta T_{Lm}}.$$

Hence, substituting for Re from (11.3.5) and taking $c_f = 0.079 \, Re^{-\frac{1}{4}}$,

$$\frac{L}{D} = 5.54 \left[\frac{D^2 \Delta p}{\nu^2 \rho} \right]^{\frac{1}{8}} \left(\frac{T_2 - T_1}{\Delta T_{Lm}} \right)^{\frac{7}{8}}, \qquad (11.3.6)$$

and this equation enables the tube length to be calculated for a heat exchanger if the temperatures and the pressure drop are specified. Practical considerations normally determine the choice of the tube diameter D. The use of a large number of small-diameter tubes will reduce the size and weight of the heat exchanger but will increase the difficulty of manufacture and the difficulty of cleaning.

11.4. Application to heat-exchanger design

The relationship between pressure drop and heat transfer, and the use of the equations derived above, can best be illustrated by a numerical example.

An air heater of the shell and tube type is to be constructed using steel tubes of 1 in. inside diameter. Air is to flow through the tubes at a total mass rate of 2 lb./sec., and is to be heated by means of steam condensing on the outside of the tubes. The tube wall temperature may be taken as 100°C. It is required to heat the air from 15 to 75°C., and the pressure drop is not to exceed 3 in. of water. Calculate the number of tubes required in parallel and their length. Assume that $c_f = 0.079 Re^{-\frac{1}{4}}$ and that for air $St = 0.046 Re^{-\frac{1}{4}}$.

Taking $\Delta p = 3$ in. water $= \frac{1}{4} \times 62 \cdot 5 \times 32 \cdot 2$ poundals/sq.ft., and the density of the air $\rho = 0 \cdot 0693$ lb./cu.ft. at 45°C., we have

$$\frac{\Delta p}{\rho} = 7 \cdot 26 \times 10^3.$$

For the temperature ratio, $T_2 - T_1 = 60^\circ$C. and $\Delta T_{Lm} = 49^\circ$C., so that

$$\frac{\Delta T_{Lm}}{T_2 - T_1} = 0 \cdot 817 \quad \text{or} \quad \frac{T_2 - T_1}{\Delta T_{Lm}} = 1 \cdot 224.$$

Hence from (11.3.4) the maximum permissible air velocity is given by

$$u_m = \sqrt{\left\{ 2 \times \frac{0 \cdot 046}{0 \cdot 079} \times 7 \cdot 26 \times 10^3 \times 0 \cdot 817 \right\}} = 83 \text{ ft./sec.}$$

The kinematic viscosity of air at 45°C. is $1 \cdot 86 \times 10^{-4}$ sq.ft./sec., and hence the Reynolds number for the air flow is

$$Re = \frac{u_m D}{\nu} = 3 \cdot 72 \times 10^4,$$

which could alternatively have been calculated directly from (11.3.5),

$$Re^{\frac{1}{4}} = 13 \cdot 89 \quad \text{and hence} \quad St = 0 \cdot 00331.$$

From (11.3.2)

$$\frac{L}{D} = \frac{1 \cdot 224}{4 \times 0 \cdot 00331} = 92 \cdot 5,$$

hence the tube length required is $L = 92 \cdot 5 \times \frac{1}{12} = 7 \cdot 7$ ft.

The number of tubes required in parallel is given by

$$N = \frac{2 \cdot 0}{0 \cdot 0693 \times 83 \times \frac{1}{4}\pi \times \frac{1}{144}} = 64 \text{ tubes.}$$

CHAPTER 12

EQUATIONS OF MOTION FOR A VISCOUS FLUID

12.1. Stresses in a viscous fluid

It has already been noted that, owing to the action of viscosity, shearing stresses can be set up in a fluid which is in motion. The stresses that can act on a fluid element $dx\, dy\, dz$ are shown in fig. 71.

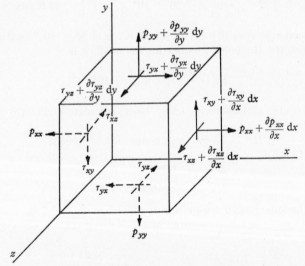

(*Note* that to avoid overcrowding the diagram stresses are omitted from the two faces perpendicular to z-axis.)

Fig. 71

The state of stress is similar to that occurring in an elastic solid. The *nine stress components* acting at a point are:

	x plane	y plane	z plane
x direction	p_{xx}	τ_{yx}	τ_{zx}
y direction	τ_{xy}	p_{yy}	τ_{zy}
z direction	τ_{xz}	τ_{yz}	p_{zz}

The notation will be clear from a comparison of this table with the diagram of fig. 71. Following the usual convention in the theory of elasticity, the normal stresses p_{xx}, p_{yy}, p_{zz} are taken as being positive in sign for tension. The components of the *resultant force per unit volume* acting on the element of fluid shown in fig. 71 are:

	x plane	y plane	z plane
x component	$\dfrac{\partial p_{xx}}{\partial x}$	$\dfrac{\partial \tau_{yx}}{\partial y}$	$\dfrac{\partial \tau_{zx}}{\partial z}$
y component	$\dfrac{\partial \tau_{xy}}{\partial x}$	$\dfrac{\partial p_{yy}}{\partial y}$	$\dfrac{\partial \tau_{zy}}{\partial z}$
z component	$\dfrac{\partial \tau_{xz}}{\partial x}$	$\dfrac{\partial \tau_{yz}}{\partial y}$	$\dfrac{\partial p_{zz}}{\partial z}$

The components of the *acceleration of the fluid element* can be written down directly from §2.3:

$$\frac{Du}{Dt} = \frac{\partial u}{\partial t} + u\frac{\partial u}{\partial x} + v\frac{\partial u}{\partial y} + w\frac{\partial u}{\partial z},$$

$$\frac{Dv}{Dt} = \frac{\partial v}{\partial t} + u\frac{\partial v}{\partial x} + v\frac{\partial v}{\partial y} + w\frac{\partial v}{\partial z},$$

$$\frac{Dw}{Dt} = \frac{\partial w}{\partial t} + u\frac{\partial w}{\partial x} + v\frac{\partial w}{\partial y} + w\frac{\partial w}{\partial z}.$$

We must allow for the possibility of a *body force* (e.g. gravity) in addition to the surface stresses shown in fig. 71. Let this be expressed *per unit mass of fluid* by the vector **F** having components F_x, F_y, F_z.

We can now write down the three equations of motion for the x, y and z directions in the following form:

$$\left.\begin{aligned}
x \text{ direction:} \quad & \rho\frac{Du}{Dt} = \rho F_x + \frac{\partial p_{xx}}{\partial x} + \frac{\partial \tau_{yx}}{\partial y} + \frac{\partial \tau_{zx}}{\partial z}, \\[2mm]
y \text{ direction:} \quad & \rho\frac{Dv}{Dt} = \rho F_y + \frac{\partial \tau_{xy}}{\partial x} + \frac{\partial p_{yy}}{\partial y} + \frac{\partial \tau_{zy}}{\partial z}, \\[2mm]
z \text{ direction:} \quad & \rho\frac{Dw}{Dt} = \rho F_z + \frac{\partial \tau_{xz}}{\partial x} + \frac{\partial \tau_{yz}}{\partial y} + \frac{\partial p_{zz}}{\partial z}.
\end{aligned}\right\} \quad (12.1.1)$$

We require further information about the stress components, however, before these equations can be of direct use.

12.2. Relationship between stress and rate of strain

In the theory of elasticity the stress components are related to the strain components in a linear manner. Similarly, in the theory of motion of viscous Newtonian fluids the components of stress are related in a linear manner to the components of the *rate of strain*. The following results are derived in Appendix 4:

For the *direct stresses*:

$$p_{xx} = -p + 2\mu \frac{\partial u}{\partial x} - \tfrac{2}{3}\mu \text{ div } \mathbf{u},$$

$$p_{yy} = -p + 2\mu \frac{\partial v}{\partial y} - \tfrac{2}{3}\mu \text{ div } \mathbf{u}, \qquad (12.2.1)$$

$$p_{zz} = -p + 2\mu \frac{\partial w}{\partial z} - \tfrac{2}{3}\mu \text{ div } \mathbf{u},$$

and for the *shearing stresses*:

$$\tau_{xy} = \tau_{yx} = \mu \left(\frac{\partial v}{\partial x} + \frac{\partial u}{\partial y} \right),$$

$$\tau_{yz} = \tau_{zy} = \mu \left(\frac{\partial w}{\partial y} + \frac{\partial v}{\partial z} \right), \qquad (12.2.2)$$

$$\tau_{zx} = \tau_{xz} = \mu \left(\frac{\partial u}{\partial z} + \frac{\partial w}{\partial x} \right).$$

Note that p in (12.2.1) is the *fluid pressure* and is positive for compression. Numerically it is equal to the mean of the three direct stresses p_{xx}, p_{yy}, p_{zz}, i.e.

$$p = -\tfrac{1}{3}(p_{xx} + p_{yy} + p_{zz}). \qquad (12.2.3)$$

The constant of proportionality μ in (12.2.1) and (12.2.2) is the *coefficient of viscosity* for the fluid.

For the special case of flow in a pipe or in a boundary layer, with significant velocities only in the x direction and with a velocity distribution which is a function of y only, the relationship (12.2.2) reduces to $\tau = \mu \, \partial u/\partial y$, which is the form most frequently encountered in engineering problems.

12.3. Navier-Stokes equation

If we substitute for the stress components from (12.2.1) and (12.2.2) in the first equation of (12.1.1) for motion in the x direction, we have

$$\rho \frac{Du}{Dt} = \rho F_x - \frac{\partial p}{\partial x} + 2\mu \frac{\partial^2 u}{\partial x^2} - \tfrac{2}{3}\mu \frac{\partial}{\partial x} (\text{div } \mathbf{u})$$

$$+ \mu \left(\frac{\partial^2 v}{\partial x \partial y} + \frac{\partial^2 u}{\partial y^2} \right) + \mu \left(\frac{\partial^2 u}{\partial z^2} + \frac{\partial^2 w}{\partial x \partial z} \right)$$

$$= \rho F_x - \frac{\partial p}{\partial x} + \mu \left(\frac{\partial^2 u}{\partial x^2} + \frac{\partial^2 u}{\partial y^2} + \frac{\partial^2 u}{\partial z^2} \right)$$
$$+ \mu \frac{\partial}{\partial x} \left(\frac{\partial u}{\partial x} + \frac{\partial v}{\partial y} + \frac{\partial w}{\partial z} \right) - \tfrac{2}{3}\mu \frac{\partial}{\partial x} (\text{div } \mathbf{u})$$

or
$$\rho \frac{D u}{D t} = \rho F_x - \frac{\partial p}{\partial x} + \mu \nabla^2 u + \tfrac{1}{3}\mu \frac{\partial}{\partial x} (\text{div } \mathbf{u}), \qquad (12.3.1)$$

and similarly for the y and z directions. These are the Navier-Stokes equations in their complete form for the motion of a viscous fluid.

The three component equations may be combined into one vectorial equation:

$$\rho \frac{D\mathbf{u}}{Dt} = \rho \mathbf{F} - \text{grad } p + \mu \nabla^2 \mathbf{u} + \frac{\mu}{3} \text{grad div } \mathbf{u}. \qquad (12.3.2)$$

For *incompressible flow* from the equation of continuity, div $\mathbf{u} = 0$, and hence in this case (12.3.2) reduces to

$$\rho \frac{D\mathbf{u}}{Dt} = \rho \mathbf{F} - \text{grad } p + \mu \nabla^2 \mathbf{u} \qquad (12.3.3)$$

or
$$\frac{D\mathbf{u}}{Dt} = \mathbf{F} - \frac{1}{\rho} \text{grad } p + \nu \nabla^2 \mathbf{u}. \qquad (12.3.3a)$$

The x component of equation $(12.3.3a)$ will be written out in full for later reference:

$$\frac{\partial u}{\partial t} + u \frac{\partial u}{\partial x} + v \frac{\partial u}{\partial y} + w \frac{\partial u}{\partial z} = F_x - \frac{1}{\rho} \frac{\partial p}{\partial x} + \nu \left(\frac{\partial^2 u}{\partial x^2} + \frac{\partial^2 u}{\partial y^2} + \frac{\partial^2 u}{\partial z^2} \right).$$

The Navier-Stokes equation applies in principle both to laminar and turbulent flow. It cannot be used directly to solve problems in turbulent flow, however, because of the impossibility of following all the minor fluctuations in velocity associated with turbulence. The Navier-Stokes equation does *not* apply to turbulent flow with mean velocities substituted in the equation in place of instantaneous values, but this problem will be discussed in Chapter 14.

If the body force $\mathbf{F}$ in the Navier-Stokes equation refers only to gravitational force, and if we are considering submerged flow with no free surface, this term may be omitted from the equation on the understanding that the local pressure p will be measured relative to the undisturbed hydrostatic pressure which would occur with equilibrium under gravity at the point considered.

12.4. Dynamical similarity

With the help of the Navier-Stokes equation of motion it will now be possible to give a rigorous proof of the principle of dynamical similarity and the use of dimensionless groups. Consider two cases of submerged flow past geometrically similar boundaries. We require to find the conditions for similarity of the flow patterns.

The equation of motion for incompressible flow in the x direction is

$$\frac{\partial u}{\partial t} + u\frac{\partial u}{\partial x} + v\frac{\partial u}{\partial y} + w\frac{\partial u}{\partial z} = -\frac{1}{\rho}\frac{\partial p}{\partial x} + \nu\left(\frac{\partial^2 u}{\partial x^2} + \frac{\partial^2 u}{\partial y^2} + \frac{\partial^2 u}{\partial z^2}\right).$$

$$(12.4.1)$$

Let U_1 be a representative velocity (e.g. the undisturbed velocity of the fluid stream in the case of flow past an obstacle, or the mean velocity for flow in a pipe). Let L be a representative length (e.g. the diameter). We can then introduce dimensionless ratios for the velocity components, the space coordinates, time and fluid pressure. These are defined as follows:

$$u' = \frac{u}{U_1}, \quad v' = \frac{v}{U_1}, \quad w' = \frac{w}{U_1},$$

$$x' = \frac{x}{L}, \quad y' = \frac{y}{L}, \quad z' = \frac{z}{L},$$

$$t' = \frac{U_1 t}{L}, \quad p' = \frac{p}{\rho U_1^2}.$$

Substituting in equation (12.4.1) will give

$$\frac{U_1^2}{L}\frac{\partial u'}{\partial t'} + \frac{U_1^2}{L}\left(u'\frac{\partial u'}{\partial x'} + v'\frac{\partial u'}{\partial y'} + w'\frac{\partial u'}{\partial z'}\right) = -\frac{1}{\rho}\frac{\rho U_1^2}{L}\frac{\partial p'}{\partial x'}$$

$$+ \frac{\nu U_1}{L^2}\left(\frac{\partial^2 u'}{\partial x'^2} + \frac{\partial^2 u'}{\partial y'^2} + \frac{\partial^2 u'}{\partial z'^2}\right),$$

i.e. $$\frac{Du'}{Dt'} = -\frac{\partial p'}{\partial x'} + \frac{\nu}{U_1 L}\left(\frac{\partial^2 u'}{\partial x'^2} + \frac{\partial^2 u'}{\partial y'^2} + \frac{\partial^2 u'}{\partial z'^2}\right). \quad (12.4.2)$$

The only term in this dimensionless equation involving the *magnitude* of the linear scale, velocity, density and viscosity is the group

$$\frac{\nu}{U_1 L} = \frac{\mu}{\rho U_1 L} = \frac{1}{Re}.$$

For given boundary conditions (fixed by the geometry of the problem), the solution of equation (12.4.2) and therefore the flow pattern depends only on the Reynolds number Re.

For the stresses p_{xx}, τ_{xy}, etc.,

$$p_{xx} = -p + 2\mu \frac{\partial u}{\partial x} = -p'\rho U_1^2 + 2\mu \frac{U_1}{L} \frac{\partial u'}{\partial x'},$$

i.e.

$$\frac{p_{xx}}{\rho U_1^2} = -p' + \frac{2}{Re} \frac{\partial u'}{\partial x'}, \qquad (12.4.3)$$

and similarly

$$\frac{\tau_{xy}}{\rho U_1^2} = \frac{1}{Re} \left(\frac{\partial v'}{\partial x'} + \frac{\partial u'}{\partial y'} \right). \qquad (12.4.4)$$

The conclusion is that the velocity ratios u/U_1, etc., the pressure coefficient $p/\rho U_1^2$, and the stress-component coefficients $p_{xx}/\rho U_1^2$, $\tau_{xy}/\rho U_1^2$, etc., are functions of x/L, y/L, z/L and Re only.

In other words, at *corresponding points* in geometrically similar systems the velocity ratios, the pressure coefficient, and the stress-component coefficients are functions of Re only.

For flow with a *free surface*, however, we must include a gravitational term. The vertical component of the equation of motion for incompressible flow, for instance, is

$$\frac{Dv}{Dt} = -g - \frac{1}{\rho} \frac{\partial p}{\partial y} + \nu \left(\frac{\partial^2 v}{\partial x^2} + \frac{\partial^2 v}{\partial y^2} + \frac{\partial^2 v}{\partial z^2} \right). \qquad (12.4.5)$$

Substituting the dimensionless ratios, as before, will give

$$\frac{U_1^2}{L} \frac{Dv'}{Dt'} = -g - \frac{U_1^2}{L} \frac{\partial p'}{\partial y'} + \frac{\nu U_1}{L^2} \left(\frac{\partial^2 v'}{\partial x'^2} + \frac{\partial^2 v'}{\partial y'^2} + \frac{\partial^2 v'}{\partial z'^2} \right),$$

i.e.

$$\frac{Dv'}{Dt'} = -\frac{gL}{U_1^2} - \frac{\partial p'}{\partial y'} + \frac{\nu}{U_1 L} \left(\frac{\partial^2 v'}{\partial x'^2} + \frac{\partial^2 v'}{\partial y'^2} + \frac{\partial^2 v'}{\partial z'^2} \right).$$

For given boundary conditions, the solution of the equation and therefore also the flow pattern will now depend on the two ratios

$$\frac{gL}{U_1^2} \quad \text{and} \quad \frac{\nu}{U_1 L}.$$

The former is the reciprocal of the *Froude number* $Fr = U_1^2/gL$, while the latter is the reciprocal of the Reynolds number again.

The physical significance of the Froude number is that it gives a measure of the ratio of inertial forces to gravity force, just as the Reynolds number gives a measure of the ratio of inertial forces to viscous forces.

12.5. Flow between parallel walls

The full equations of motion are of limited use to the engineer because of the mathematical difficulty of finding solutions. A few simple examples will be given, however, in the following section to illustrate the physical meaning of the equations.

The simplest case of engineering interest is that of laminar flow of a viscous liquid between parallel walls. The problem is represented diagrammatically in fig. 72. The equations of motion in two dimensions are

$$u \frac{\partial u}{\partial x} + v \frac{\partial u}{\partial y} = -\frac{1}{\rho} \frac{\partial p}{\partial x} + \nu \left(\frac{\partial^2 u}{\partial x^2} + \frac{\partial^2 u}{\partial y^2} \right), \qquad (12.5.1)$$

$$u \frac{\partial v}{\partial x} + v \frac{\partial v}{\partial y} = -\frac{1}{\rho} \frac{\partial p}{\partial y} + \nu \left(\frac{\partial^2 v}{\partial x^2} + \frac{\partial^2 v}{\partial y^2} \right), \qquad (12.5.2)$$

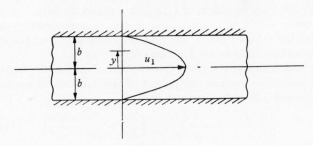

Fig. 72

and the continuity equation is

$$\frac{\partial u}{\partial x} + \frac{\partial v}{\partial y} = 0. \qquad (12.5.3)$$

For final steady conditions with *fully developed flow* $\partial u/\partial x = 0$, i.e. $u = f(y)$ only, and the transverse velocity component v must be zero everywhere. The equations of motion therefore reduce to

$$-\frac{1}{\rho} \frac{\partial p}{\partial x} + \nu \frac{\partial^2 u}{\partial y^2} = 0 \qquad (12.5.4)$$

and

$$-\frac{1}{\rho} \frac{\partial p}{\partial y} = 0. \qquad (12.5.5)$$

From (12.5.5) the pressure p must be constant across any section perpendicular to the flow, i.e. $p = f(x)$ only. But from (12.5.4) $\frac{\partial p}{\partial x} = \mu \frac{\partial^2 u}{\partial y^2}$ and therefore, since u is a function of y only, the pressure gradient $\partial p/\partial x$ must be constant. Hence

$$\frac{\partial p}{\partial x} = \text{constant} = -\frac{\Delta p}{L} \text{ say,}$$

and therefore, from (12.5.4),

$$\frac{\mathrm{d}^2u}{\mathrm{d}y^2} = -\frac{\Delta p}{\mu L}. \tag{12.5.6}$$

Integrating, and noting that $\mathrm{d}u/\mathrm{d}y = 0$ at the centre of the channel where $y = 0$,

$$\frac{\mathrm{d}u}{\mathrm{d}y} = -\frac{\Delta p}{\mu L}\, y,$$

and

$$u = -\frac{\Delta p}{2\mu L}\, y^2 + B.$$

But at $y = b$ the velocity $u = 0$, and therefore the constant of integration

$$B = \frac{\Delta p}{2\mu L}\, b^2.$$

Hence,

$$u = \frac{\Delta p}{2\mu L}\, (b^2 - y^2). \tag{12.5.7}$$

Taking unit breadth in the z direction, the total flow is given by

$$Q = \int_{-b}^{+b} u\, \mathrm{d}y = \frac{2}{3}\frac{\Delta p}{\mu L}\, b^3, \tag{12.5.8}$$

and therefore the mean velocity, averaged across the channel, is

$$u_m = \frac{Q}{2b} = \tfrac{2}{3}u_1. \tag{12.5.9}$$

Compare these results with those for laminar flow in a circular pipe given by (6.2.2), (6.2.3) and (6.2.4).

12.6. Laminar flow of a viscous liquid film down a vertical wall

Another simple case of viscous flow is that of a liquid film on a vertical wall. Taking the x direction vertically downwards, the equation of motion is now

$$u\,\frac{\partial u}{\partial x} + v\,\frac{\partial u}{\partial y} = -\frac{1}{\rho}\frac{\partial p}{\partial x} + g + \nu\left(\frac{\partial^2 u}{\partial x^2} + \frac{\partial^2 u}{\partial y^2}\right), \tag{12.6.1}$$

and for the final steady velocity profile $\partial u/\partial x = 0$, $\partial^2 u/\partial x^2 = 0$, $v = 0$, and because there is a free surface $\partial p/\partial x = 0$, and hence equation (12.6.1) reduces to

$$g + \nu\,\frac{\partial^2 u}{\partial y^2} = 0. \tag{12.6.2}$$

Integrating,
$$\frac{du}{dy} = -\frac{g}{\nu}y + A,$$

and when $y = \delta$, $du/dy = 0$, since the shear stress must be zero at the surface of the film. Therefore

$$\frac{du}{dy} = \frac{g}{\nu}(\delta - y),$$

and integrating again,

$$u = \frac{g}{\nu}\left(\delta y - \frac{y^2}{2}\right) + B.$$

But when $y = 0$, $u = 0$ for zero slip at the wall, i.e. $B = 0$ and the velocity distribution is given by

$$u = \frac{g}{\nu}\left(\delta y - \frac{y^2}{2}\right). \tag{12.6.3}$$

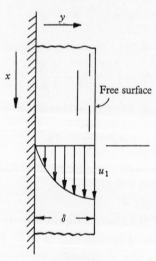

Fig. 73

The volume flow per unit width is given by

$$Q = \int_0^\delta u \, dy = \frac{g}{\nu}\frac{\delta^3}{3} \tag{12.6.4}$$

or, alternatively, the final thickness of the film is given by

$$\delta = \left[\frac{3\nu Q}{g}\right]^{\frac{1}{3}}. \tag{12.6.5}$$

12.7. Flow past a sphere

For incompressible flow with zero body force acting on the fluid, the Navier-Stokes equation of motion may be written

$$\rho \frac{\partial \mathbf{u}}{\partial t} + \rho \mathbf{u} \,.\, \nabla \mathbf{u} = - \operatorname{grad} p + \mu \nabla^2 \mathbf{u}, \tag{12.7.1}$$

and for steady flow the equation becomes

$$\rho \mathbf{u} \,.\, \nabla \mathbf{u} = - \operatorname{grad} p + \mu \nabla^2 \mathbf{u}. \tag{12.7.2}$$

This equation is intractable, however, when applied to the problem of flow past a sphere with the necessary boundary condition that the velocity vector $\mathbf{u} = 0$ at all points on the surface of the sphere.

For flow at low velocity or *small values of the Reynolds number*, however, it would be reasonable to neglect the inertial terms on the left-hand side of equation (12.7.2). The equation to be solved in this case is

$$\mu \nabla^2 \mathbf{u} = \operatorname{grad} p. \tag{12.7.3}$$

Stokes's solution may be obtained by expressing this equation, together with the continuity equation, in spherical polar coordinates and fitting the boundary conditions that the velocity components are zero at the surface of the sphere where $r = a$. The pressure distribution over the surface can then be calculated and the total drag is found to be

$$D = 6\pi\mu a u_1, \tag{12.7.4}$$

where a is the radius of the sphere, and u_1 is the undisturbed velocity of the stream.

If the result is expressed in terms of a drag coefficient c_D

$$c_D = \frac{D}{\frac{1}{2}\rho u_1^2 \pi a^2} = \frac{12\mu}{\rho u_1 a}$$

or

$$c_D = \frac{24}{Re}, \tag{12.7.5}$$

where the Reynolds number Re is defined by $Re = \rho u_1 d / \mu$. For the mathematical details of Stokes's solution the reader should consult [1].

The result given above for the drag of a sphere is found to be in good agreement with the measured drag for values of the Reynolds number less than about 2. At larger values of the Reynolds number separation of the flow occurs on the downstream side of the sphere with the formation of an unsteady eddying wake. At still higher values of the Reynolds number the flow in the wake becomes completely turbulent. It is no longer possible in these circumstances to calculate the drag from the Navier-Stokes equation and recourse must be had to direct measurement. The nature of the flow is shown diagrammatically in fig. 74 and the drag coefficient is plotted against Reynolds number in fig. 75.

It is possible to divide the drag into two components, skin-friction drag and form drag. The *skin-friction drag* is given by the integral of the tangential shearing stress taken over the entire surface of the sphere. The *form drag* is given by the integral of the normal stress component taken over the surface. Separation of the flow, with the formation of an eddying or turbulent wake as illustrated in fig. 74, causes failure to achieve the full pressure recovery over the rear half of the sphere, and this contributes directly to the form drag.

At large values of the Reynolds number, the flow over the front half of a sphere may be divided into a thin boundary-layer region

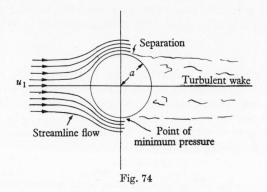

Fig. 74

in which the effects of viscosity are important, and an outer region in which the flow corresponds to that of an inviscid fluid. The normal pressure distribution over the front half may then be calculated from the theory of flow of ideal or non-viscous fluids, and the tangential shearing stress can be calculated approximately from boundary-layer theory. This method of analysis, however, cannot be extended to flow over the rear half of the sphere.

Over the front half of the sphere the fluid experiences a falling pressure from the front stagnation point to the point of minimum pressure. This favourable pressure gradient has a stabilizing effect on the boundary layer which therefore remains laminar. Once the minimum pressure point is passed, however, half-way round the surface of the sphere, the boundary layer has to face an adverse pressure gradient. For reasons which will be explained in Chapter 13, this leads rapidly to separation of the boundary layer and the formation of a wake. Laminar separation generally occurs at only a very short distance behind the minimum pressure point. At very large values of the Reynolds number, however, transition to turbulent flow in the boundary layer will take place before laminar separation can occur. In these circumstances the boundary layer with turbulent flow will adhere to the surface for a slightly greater distance before turbulent separation occurs. The effect of this change in the flow is a net decrease in the total drag due to the smaller form drag. The sudden change in the drag coefficient is shown on fig. 75, occurring at a value of Re between 10^5 and 10^6.

At large values of the Reynolds number the total drag of a sphere, or of any other bluff-shaped body, is mainly due to form drag. The

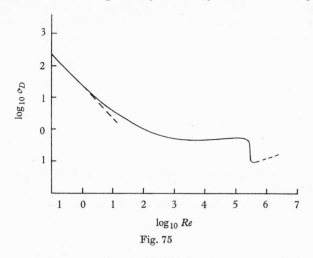

Fig. 75

object of *streamlining* is the reduction of form drag. This can be achieved by designing shapes with long tapering extensions on the downstream side to avoid the occurrence of large adverse pressure gradients. Inevitably there will be some increase in the skin-friction drag owing to the larger surface area and to the fact that, beyond the point of minimum pressure, flow in the boundary layer itself is liable to be turbulent with correspondingly greater shearing stresses. However, the reduction of form drag by streamlining will generally outweigh any increase in skin friction and will thus achieve a substantial reduction in the total drag.

REFERENCES

(1)   LAMB. *Hydrodynamics* (Cambridge University Press).
(2)   GOLDSTEIN. *Modern Developments in Fluid Dynamics* (Oxford).

CHAPTER 13

BOUNDARY LAYERS

13.1. The boundary-layer equations for laminar flow

Consider the flow of a uniform stream, with undisturbed velocity u_1, past a flat surface as indicated in fig. 76. Let x be the distance measured from the leading edge of the plate or surface in the direction of flow of the undisturbed stream. Let y be the distance measured perpendicular to the surface. The problem may be treated

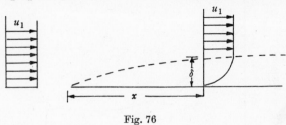

Fig. 76

as a two-dimensional one, i.e. velocity components in the z direction will be ignored.

The full equations of motion in two dimensions are:

$$u \frac{\partial u}{\partial x} + v \frac{\partial u}{\partial y} = -\frac{1}{\rho} \frac{\partial p}{\partial x} + \nu \left[\frac{\partial^2 u}{\partial x^2} + \frac{\partial^2 u}{\partial y^2} \right], \qquad (13.1.1)$$

$$u \frac{\partial v}{\partial x} + v \frac{\partial v}{\partial y} = -\frac{1}{\rho} \frac{\partial p}{\partial y} + \nu \left[\frac{\partial^2 v}{\partial x^2} + \frac{\partial^2 v}{\partial y^2} \right], \qquad (13.1.2)$$

and $$\frac{\partial u}{\partial x} + \frac{\partial v}{\partial y} = 0. \qquad (13.1.3)$$

The first two equations, (13.1.1) and (13.1.2), are the Navier-Stokes equations in two-dimensional form for steady incompressible flow, and (13.1.3) is the equation of continuity.

The boundary conditions which must be satisfied are that at the surface $y = 0$, $u = 0$ and $v = 0$, where u and v are the velocity components in the x and y directions.

It is found experimentally with flow at large Reynolds number that the boundary-layer region, in which the velocity varies from zero at the wall to the full value u_1 for the undisturbed stream, is relatively thin. This may be expressed mathematically by saying that, if δ is the thickness of the boundary layer at distance x from the leading edge of the plate, δ *is small compared with* x. With the help of this experimental observation we can now proceed to simplify equations (13.1.1) and (13.1.2).

Let x be regarded as the standard order of size for distances, and let u_1 be regarded as the standard order of size for velocities. The x component of velocity u varies from zero at $y = 0$ to u_1 at $y = \delta$. We can therefore write the orders of magnitude of the velocity component u and its derivatives as follows:

$$u = O(1), \quad \frac{\partial u}{\partial y} = O\left(\frac{1}{\delta}\right), \quad \frac{\partial^2 u}{\partial y^2} = O\left(\frac{1}{\delta^2}\right),$$

$$\frac{\partial u}{\partial x} = O(1), \quad \frac{\partial^2 u}{\partial x^2} = O(1).$$

Also from the equation of continuity (13.1.3) $\partial v/\partial y$ cannot be greater than order (1), and since the range of y considered is from 0 to δ, the velocity component v must be of order δ. The following statements can therefore be made:

$$v = O(\delta), \quad \frac{\partial v}{\partial y} = O(1), \quad \frac{\partial^2 v}{\partial y^2} = O\left(\frac{1}{\delta}\right),$$

$$\frac{\partial v}{\partial x} = O(\delta), \quad \frac{\partial^2 v}{\partial x^2} = O(\delta).$$

If equation (13.1.1) is now examined, it will be evident that on the right-hand side the term $\partial^2 u/\partial x^2$ may be neglected in comparison with $\partial^2 u/\partial y^2$. The equation therefore becomes

$$u \frac{\partial u}{\partial x} + v \frac{\partial u}{\partial y} = -\frac{1}{\rho} \frac{\partial p}{\partial x} + \nu \frac{\partial^2 u}{\partial y^2}, \tag{13.1.4}$$

and if the term $\nu(\partial^2 u/\partial y^2)$ describing the viscous force is assumed to be of the same order of magnitude as the inertia terms on the left-hand side of the equation we can say that ν must be $O(\delta^2)$, i.e.

$$\delta \quad \text{is} \quad O(\nu^{\frac{1}{2}}).$$

Now examine equation (13.1.2). Apart from $\dfrac{1}{\rho} \dfrac{\partial p}{\partial y}$, all the terms in the equation are $O(\delta)$ or smaller. The term $\dfrac{1}{\rho} \dfrac{\partial p}{\partial y}$ cannot therefore be larger than $O(\delta)$. In other words the total change of pressure in a transverse direction through the boundary layer is $O(\delta^2)$ and may therefore be neglected. Equation (13.1.2) can thus be reduced to

$$\frac{\partial p}{\partial y} = 0. \tag{13.1.5}$$

The Reynolds number for the flow, specified with the distance x measured from the leading edge, is $Re = u_1 x/v$, and this must be $O(1/v)$ in magnitude. Therefore, since v is $O(\delta^2)$ we can say

$$\delta = O(Re^{-\frac{1}{2}}) \quad \text{or} \quad \frac{\delta}{x} = O(Re^{-\frac{1}{2}}). \tag{13.1.6}$$

This is in accordance with the earlier observation that, the greater the value of the Reynolds number, the smaller will be the relative magnitude of the viscous forces compared with inertia forces, and the greater will be the tendency for viscous effects to be confined to a relatively thin layer adjacent to the solid wall or boundary of the flow. We could interpret (13.1.6) in a slightly different way by saying that the *condition* for the existence of a thin boundary-layer region is that the Reynolds number should be relatively large.

The boundary-layer equations (13.1.4) and (13.1.5) may be solved for the case of flow past a flat plate with zero pressure gradient $\partial p/\partial x$ in the x direction by assuming that the velocity profiles are similar, i.e. that $u/u_1 = f(y/\delta)$ only, and by anticipating the result that the boundary-layer thickness δ is actually proportional to $x^{\frac{1}{2}}$, i.e. that $\delta/x \sim Re^{-\frac{1}{2}}$. The exact solution is given in Appendix 5.

13.2. Boundary-layer thickness

Difficulty arises in specifying the boundary-layer thickness because the velocity profile must merge imperceptibly into the main

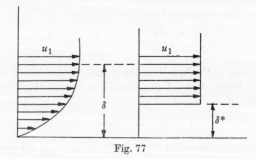

Fig. 77

stream outside the boundary layer, i.e. as $y \to \delta$, $u \to u_1$ and $\partial u/\partial y \to 0$.

A quantity known as the *displacement thickness* δ^* can be defined in a precise way, however, as follows. Referring to fig. 77, let the volume flow in the boundary layer be Q. Then

$$Q = \int_0^\delta u \, \mathrm{d}y = u_1(\delta - \delta^*),$$

i.e.
$$\delta^* = \frac{1}{u_1} \int_0^\delta (u_1 - u)\, dy$$

or
$$\delta^* = \int_0^\delta \left(1 - \frac{u}{u_1}\right) dy. \qquad (13.2.1)$$

The physical meaning of this definition is that δ^* represents the distance by which an equivalent uniform stream would have to be

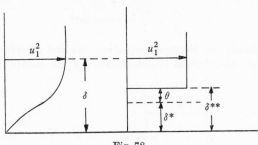

Fig. 78

displaced from the surface, as shown in fig. 77, to give the same volume flow.

A similar picture may be drawn for the momentum flow in the boundary layer. Referring to fig. 78, let the momentum flow be M. Then
$$M = \int_0^\delta \rho u^2\, dy = \rho u_1^2 (\delta - \delta^{**}),$$

i.e. assuming constant density,
$$\delta^{**} = \int_0^\delta \left(1 - \frac{u^2}{u_1^2}\right) dy. \qquad (13.2.2)$$

It would be logical to describe δ^{**} as the momentum thickness, but in practice this name is reserved for another quantity θ defined by $\theta = \delta^{**} - \delta^*$ as shown in fig. 78. Substituting from (13.2.1) and (13.2.2), the *momentum thickness* is defined by
$$\theta = \int_0^\delta \frac{u}{u_1} \left(1 - \frac{u}{u_1}\right) dy. \qquad (13.2.3)$$

13.3. Separation and transition

Outside the boundary layer the viscous terms in the equations of motion may be neglected altogether. If we treat the flow as a one-dimensional problem, i.e. with the velocity u_1 outside the boundary layer being independent of y but possibly varying slowly with distance x measured along the surface, the equation of motion is simply

$$u_1 \frac{du_1}{dx} = -\frac{1}{\rho} \frac{dp}{dx}. \tag{13.3.1}$$

This would apply to flow past a flat surface with a pressure gradient in the x direction, or to flow past a slightly curved surface. Integrating with respect to x gives

$$\frac{p_1}{\rho} + \frac{u_1^2}{2} = \text{constant}, \tag{13.3.2}$$

which is simply the Bernoulli equation. We can therefore say that the pressure distribution is determined by the nature of the flow

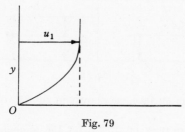

Fig. 79

outside the boundary layer and that the pressure is transmitted without change through the boundary layer to the surface.

At the outer edge of the boundary layer, where $y = \delta$, we must have $u = u_1$, $\partial u/\partial y = 0$ and $\partial^2 u/\partial y^2 = 0$.

At the wall, where $y = 0$, we must have $u = 0$ and $v = 0$, and the skin friction or shearing stress will be given by

$$\tau_0 = \mu \left(\frac{\partial u}{\partial y}\right)_0. \tag{13.3.3}$$

Also, from the boundary-layer equation (13.1.4), at the wall

$$-\frac{1}{\rho} \frac{\partial p}{\partial x} + v \left(\frac{\partial^2 u}{\partial y^2}\right)_0 = 0$$

or

$$\mu \left(\frac{\partial^2 u}{\partial y^2}\right)_0 = \frac{\partial p}{\partial x}. \tag{13.3.4}$$

In the case of *zero external pressure gradient*, it follows from (13.3.4) that $(\partial^2 u/\partial y^2)_0 = 0$ and therefore the velocity gradient $\partial u/\partial y$ must be greatest at the wall and will fall off steadily to zero at the outer edge. The velocity profile must therefore be of the general form shown in fig. 79. If, however, $\partial p/\partial x > 0$, i.e. if we have an *adverse pressure gradient*, it follows from (13.3.4) that $(\partial^2 u/\partial y^2)_0$ will be positive and therefore the velocity gradient $\partial u/\partial y$ must first of all increase with distance from the wall before it can start falling off to zero. In this case the velocity profile must be of the form shown in fig. 80 (a).

Under extreme conditions with an adverse pressure gradient the velocity profile may become increasingly distorted until the velocity gradient at the wall $(\partial u/\partial y)_0$ is zero, as indicated in fig. 80 (b). At this point *separation of flow* from the wall is said to occur. There will then be a dead-water or back-flow region adjacent to the wall downstream from the point of separation.

It is observed that the type of velocity profile shown in fig. 80 is inherently unstable, and it frequently happens that transition to turbulent flow in the boundary layer will take place before laminar

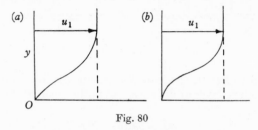

Fig. 80

separation can occur. In these circumstances the boundary layer will adhere to the surface for a greater distance with turbulent flow being maintained against the adverse pressure gradient. Eventually, however, separation of the turbulent boundary layer will occur.

Transition to turbulent flow in a boundary layer is liable to take place even if there is no adverse pressure gradient. The phenomenon of transition is extremely complex and there is no simple criterion. Turbulence can always be caused by roughness or large disturbances. It appears, however, that under certain circumstances small disturbances in a laminar boundary layer can become amplified until turbulence is developed. If we specify a Reynolds number using the displacement thickness δ^* as the characteristic length, i.e. if $Re = u_1 \delta^*/\nu$, the flow will usually remain laminar provided $Re < 600$, but for larger values of the Reynolds number the flow becomes unstable with respect to disturbances whose wavelengths fall within a certain range. This occurrence of instability is found in practice to lead rapidly into transition. For further details see [2].

13.4. The momentum equation for the boundary layer

Consider a short length dx of the boundary layer, as shown in fig. 81. Let the thickness of the boundary layer at distance x from the leading edge be δ. It is assumed for the purposes of the following discussion that δ is a finite and distinguishable quantity defined by the condition that $u = u_1$ when $y = \delta$. If, however, the velocity profile involves an asymptotic approach to the main stream velocity u_1 we can adopt some alternative definition, such as the condition that the velocity u should be within 1 per cent of the main stream velocity u_1 when $y = \delta$. So long as we keep to the same definition throughout it does not matter in which particular way δ is defined.

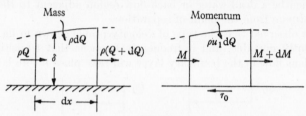

Fig. 81

Referring to fig. 81, the *mass flow* ρQ and the *momentum flow* M in the boundary layer at distance x from the leading edge are given by

$$\rho Q = \rho u_1(\delta - \delta^*) \tag{13.4.1}$$

and

$$M = \rho u_1^2(\delta - \delta^* - \theta). \tag{13.4.2}$$

These follow directly from the definitions of displacement thickness and momentum thickness given in §13.2.

If the volume flow Q included within the boundary-layer region increases by the amount dQ in distance dx, the principle of continuity requires that an inward flow dQ should occur across the outer edge of the boundary layer as shown in fig. 81. This is simply another way of expressing the fact that the boundary-layer thickness δ is increasing as the flow proceeds along the surface in the x direction. The inward volume flow dQ across the outer edge of the boundary layer involves a corresponding inward flow of momentum $\rho u_1\, dQ$. If we now take a *control surface* extending from the wall at $y = 0$ to the outer edge of the boundary layer and from distance x to $x + dx$, and if we consider unit width of flow in a direction perpendicular to the plane of the diagram, we can express the momentum equation as follows:

$$- \tau_0\, dx - \delta \frac{\partial p}{\partial x}\, dx = \text{net outflow of } x \text{ momentum}$$

$$= \frac{dM}{dx}\, dx - \rho u_1 \frac{dQ}{dx}\, dx.$$

$$\therefore \quad -\frac{\tau_0}{\rho} - \frac{\delta}{\rho}\frac{\partial p}{\partial x} = \frac{d}{dx}[u_1^2(\delta - \delta^* - \theta)] - u_1\frac{d}{dx}[u_1(\delta - \delta^*)]$$

$$= u_1^2\frac{d}{dx}(\delta - \delta^* - \theta) + 2(\delta - \delta^* - \theta)u_1\frac{du_1}{dx}$$

$$- u_1^2\frac{d}{dx}(\delta - \delta^*) - (\delta - \delta^*)u_1\frac{du_1}{dx}$$

$$= -u_1^2\frac{d\theta}{dx} + (\delta - \delta^* - 2\theta)u_1\frac{du_1}{dx},$$

but from (13.3.1)

$$u_1\frac{du_1}{dx} = -\frac{1}{\rho}\frac{dp}{dx}.$$

$$\therefore \quad -\frac{\tau_0}{\rho} = -u_1^2\frac{d\theta}{dx} - (\delta^* + 2\theta)u_1\frac{du_1}{dx}$$

or

$$\frac{\tau_0}{\rho u_1^2} = \frac{d\theta}{dx} + (2\theta + \delta^*)\frac{1}{u_1}\frac{du_1}{dx}, \tag{13.4.3}$$

which is the *momentum equation* for the boundary layer. If the pressure outside the boundary layer is constant, $du_1/dx = 0$, and the momentum equation then takes the simple form

$$\frac{\tau_0}{\rho u_1^2} = \frac{d\theta}{dx}. \tag{13.4.4}$$

13.5. Approximate solution for laminar boundary-layer flow using the momentum equation

Taking the case of zero external pressure gradient, we may assume similarity of velocity profiles, i.e. that

$$\frac{u}{u_1} = f\left(\frac{y}{\delta}\right), \tag{13.5.1}$$

where δ is the boundary-layer thickness which will be a function of x.

If we now choose a particular form for the profile of (13.5.1) we can evaluate the momentum thickness θ in terms of δ, and we can also express τ_0 in terms of δ by making use of the fact that the shear stress at the wall is given by

$$\tau_0 = \mu\left(\frac{\partial u}{\partial y}\right)_0. \tag{13.5.2}$$

By substituting for τ_0 and θ in the momentum equation (13.4.4) we can thus obtain the variation of the boundary-layer thickness with x.

The boundary conditions which should be satisfied by the velocity profile of (13.5.1) are as follows:

at $\quad y = 0 \quad u = 0 \quad \partial u/\partial y$ must be finite and $\quad \partial^2 u/\partial y^2 = 0$,

at $\quad y = \delta \quad u = u_1 \quad\quad \partial u/\partial y = 0 \quad\quad\quad \partial^2 u/\partial y^2 = 0$.

The simplest mathematical form for (13.5.1) which will meet all these requirements is

$$\frac{u}{u_1} = 2\frac{y}{\delta} - 2\left(\frac{y}{\delta}\right)^3 + \left(\frac{y}{\delta}\right)^4. \tag{13.5.3}$$

With this profile the displacement and momentum thicknesses will be given by

$$\frac{\delta^*}{\delta} = \int_0^1 \left(1 - \frac{u}{u_1}\right) \mathrm{d}\left(\frac{y}{\delta}\right) = \frac{3}{10}, \tag{13.5.4}$$

and

$$\frac{\theta}{\delta} = \int_0^1 \frac{u}{u_1}\left(1 - \frac{u}{u_1}\right) \mathrm{d}\left(\frac{y}{\delta}\right) = \frac{37}{315}, \tag{13.5.5}$$

and the shear stress at the wall will be given by

$$\tau_0 = \frac{2\mu u_1}{\delta}. \tag{13.5.6}$$

If we now substitute from (13.5.5) and (13.5.6) in the momentum equation (13.4.4), we have the following result:

$$\frac{2\nu}{u_1\delta} = \frac{37}{315}\frac{\mathrm{d}\delta}{\mathrm{d}x},$$

and hence, integrating from 0 to x,

$$\delta^2 = 34\cdot0\,\frac{\nu x}{u_1},$$

i.e. $$\delta = 5\cdot83\left(\frac{\nu x}{u_1}\right)^{\frac{1}{2}}, \tag{13.5.7}$$

and from (13.5.4) and (13.5.5)

$$\delta^* = 1\cdot75\left(\frac{\nu x}{u_1}\right)^{\frac{1}{2}}, \tag{13.5.8}$$

and $$\theta = 0\cdot686\left(\frac{\nu x}{u_1}\right)^{\frac{1}{2}}. \tag{13.5.9}$$

Equation (13.5.7) for the thickness δ may be expressed alternatively

$$\frac{\delta}{x} = 5\cdot83\left(\frac{\nu}{u_1 x}\right)^{\frac{1}{2}} = 5\cdot83\,Re^{-\frac{1}{2}}. \tag{13.5.10}$$

The skin-friction coefficient may be obtained either from (13.5.6) or directly from the momentum equation (13.4.4) with the result

$$c_f = \frac{\tau_0}{\frac{1}{2}\rho u_1^2} = 0.686 \left(\frac{u_1 x}{\nu}\right)^{-\frac{1}{2}}. \qquad (13.5.11)$$

If these results are compared with those obtained in Appendix 5 for the exact solution, it will be seen that the agreement is quite close. No value is obtained for δ in the exact solution because the approach to the undisturbed stream velocity u_1 is actually asymptotic. The displacement and momentum thicknesses, however, are in good agreement in each case.

The *drag coefficient* for a length x (or the *average* skin-friction coefficient) is given by $c_D = \dfrac{D}{\frac{1}{2}\rho u_1^2 x}$, where the drag D exerted on one side of the plate is $\displaystyle\int_0^x \tau_0 \, dx$. Hence from (13.5.11)

$$c_D = 1.372 \left(\frac{\nu}{u_1 x}\right)^{\frac{1}{2}} = 1.372 Re^{-\frac{1}{2}}, \qquad (13.5.12)$$

which should be compared with the exact result derived in Appendix 5.

Other forms for the velocity profile, which may be used in place of (13.5.3) for a similar calculation by the method of the momentum equation, are

$$\frac{u}{u_1} = 2\frac{y}{\delta} - \left(\frac{y}{\delta}\right)^2,$$

and

$$\frac{u}{u_1} = \sin\frac{\pi}{2}\frac{y}{\delta},$$

and these again give reasonably good agreement with the exact solution obtained in Appendix 5, although they do not satisfy the requirement that $\partial^2 u/\partial y^2$ should be zero when $y = \delta$ and the quadratic form does not even satisfy the condition that $\partial^2 u/\partial y^2 = 0$ at $y = 0$.

It is possible to obtain an approximate solution for laminar boundary-layer flow with an external pressure gradient in the x direction, using essentially the same method. In this case the momentum equation must be used in the complete form of (13.4.3). The problem is complicated by the fact that we can no longer assume similarity of velocity profiles, and it is therefore necessary to compound two different basic types of profile in varying proportions in order to secure an approximate representation of the actual velocity distribution. For details of the method the reader should consult [1].

13.6. Approximate method for turbulent boundary-layer flow using the momentum equation

The momentum equation may also be applied to turbulent boundary-layer flow, but we can no longer calculate the shear stress at the wall from the expression $\mu(\partial u/\partial y)_0$. It is true that the stress must ultimately be transmitted to the wall through the mechanism of viscosity in the laminar sub-layer, but we cannot hope to represent the thickness or the velocity distribution in the laminar sub-layer with complete accuracy. We therefore choose a simple velocity profile to represent the distribution in the turbulent zone outside

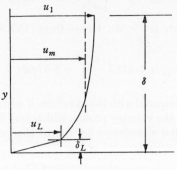

Fig. 82

the laminar sub-layer, and we make use of measurements on turbulent flow in pipes to give an empirical expression for the shear stress τ_0.

We can use the same simplified picture for the flow that was employed for the Taylor-Prandtl analogy in §11.2. It is assumed, as represented in fig. 82, that a sharp distinction can be made between the laminar sub-layer and the outer turbulent zone. For the velocity distribution in the turbulent zone we can take, as a first approximation, the one-seventh power law derived for pipe flow in §6.3, i.e.

$$\frac{u}{u_1} = \left(\frac{y}{\delta}\right)^{\frac{1}{7}}. \tag{13.6.1}$$

For the laminar sub-layer, however, a straight-line velocity distribution may be assumed, and this implies that the turbulent shear stress at the outer edge of the sub-layer is transmitted without change through the laminar region to the wall.

For the turbulent shearing stress τ_0 we can use the empirical formula which is based on pipe-flow measurements

$$\tau_0 = 0.0395\rho u_m^2 \left(\frac{u_m d}{\nu}\right)^{-\frac{1}{4}},$$

where d is the pipe diameter and u_m the mean velocity. Noting that the velocity distribution for turbulent flow in smooth pipes can be represented approximately by $u/u_1 = (y/\frac{1}{2}d)^{\frac{1}{7}}$, which corresponds with equation (13.6.1) for the boundary layer, and that the mean velocity u_m for this distribution is related to the maximum velocity u_1 by $u_m = 0.817 u_1$, we can rewrite the expression for τ_0 in a form which is directly applicable to boundary-layer flow

$$\tau_0 = 0.0233\rho u_1^2 \left(\frac{u_1 \delta}{\nu}\right)^{-\frac{1}{4}}. \tag{13.6.2}$$

The momentum thickness θ can be evaluated in terms of δ directly from (13.6.1). The departure of the velocity profile from this form in the laminar sub-layer can safely be neglected in evaluating θ. The result is that

$$\frac{\theta}{\delta} = \frac{7}{72}. \tag{13.6.3}$$

Substituting from (13.6.2) and (13.6.3) in the momentum equation (13.4.4) gives

$$0.0233 \left(\frac{u_1 \delta}{\nu}\right)^{-\frac{1}{4}} = \frac{7}{72}\frac{\mathrm{d}\delta}{\mathrm{d}x}$$

or

$$\frac{\mathrm{d}\delta}{\mathrm{d}x} = 0.239 \left(\frac{\nu}{u_1 \delta}\right)^{\frac{1}{4}},$$

and hence, integrating from 0 to x,

$$\tfrac{4}{5}\delta^{\frac{5}{4}} = 0.239 \left(\frac{\nu}{u_1}\right)^{\frac{1}{4}} x,$$

i.e.

$$\frac{\delta}{x} = 0.379 \left(\frac{\nu}{u_1 x}\right)^{\frac{1}{5}} = 0.379 Re^{-\frac{1}{5}}. \tag{13.6.4}$$

For the turbulent skin-friction coefficient, from the momentum equation, we have

$$c_f = \frac{\tau_0}{\frac{1}{2}\rho u_1^2} = 0.0588 \left(\frac{u_1 x}{\nu}\right)^{-\frac{1}{5}}, \tag{13.6.5}$$

and for the drag coefficient, or average skin-friction coefficient for one side of a plate of length x, the result is

$$c_D = 0.073 Re^{-\frac{1}{5}}. \tag{13.6.6}$$

A rough estimate may be obtained for the *thickness of the laminar sub-layer* in the following way. Assuming a linear velocity distribution in the sub-layer, as indicated in fig. 82, we will have for values of y less than δ_L

$$\frac{u}{u_L} = \frac{y}{\delta_L},\tag{13.6.7}$$

and the shearing stress τ_0 will be given by

$$\tau_0 = \mu\,\frac{u_L}{\delta_L}.\tag{13.6.8}$$

For values of y greater than δ_L the velocity distribution will be given by (13.6.1), and the turbulent shearing stress τ_0 is given by (13.6.2).

Eliminating τ_0 between (13.6.2) and (13.6.8) gives

$$\mu\,\frac{u_L}{\delta_L} = 0\cdot 0233\,\rho u_1^2 \left(\frac{u_1\delta}{\nu}\right)^{-\frac14}$$

or

$$\frac{u_L}{u_1} = 0\cdot 0233\,\frac{\delta_L}{\delta}\left(\frac{u_1\delta}{\nu}\right)^{\frac34}.$$

And fitting the two velocity profiles together to give $u = u_L$ at $y = \delta_L$, we must have

$$\frac{u_L}{u_1} = \left(\frac{\delta_L}{\delta}\right)^{\frac17}.$$

Hence,

$$\frac{u_L}{u_1} = 1\cdot 87\left(\frac{u_1\delta}{\nu}\right)^{-\frac18}\tag{13.6.9}$$

and

$$\frac{\delta_L}{\delta} = 80\left(\frac{u_1\delta}{\nu}\right)^{-\frac78}.\tag{13.6.10}$$

Also from (13.6.2) we can say

$$\sqrt{\frac{\tau_0}{\rho}} = 0\cdot 152\,u_1\left(\frac{u_1\delta}{\nu}\right)^{-\frac18}.\tag{13.6.11}$$

Hence from (13.6.10) and (13.6.11)

$$\frac{\delta_L\sqrt{(\tau_0/\rho)}}{\nu} = 12\tag{13.6.12}$$

or

$$\delta_L = \frac{12\nu}{\sqrt{(\tau_0/\rho)}},$$

which confirms the statement made in §6.3 regarding the order of magnitude of the thickness of the laminar sub-layer.

Further discussion of turbulent boundary-layer flow is reserved for Chapter 14.

REFERENCES

(1)   GOLDSTEIN. *Modern Developments in Fluid Dynamics* (Oxford).
(2)   SCHUBAUER and SKRAMSTAD. *J. Aero. Sci.* **14** (1947).

CHAPTER 14

TURBULENT FLOW

14.1. Fluctuating velocity components and the Reynolds stresses

The Navier-Stokes equations would cover the case of turbulent flow if we could follow out all the minor fluctuations in velocity, and turbulence would then be regarded as an extreme case of unsteady motion. Such a procedure, however, would obviously be impracticable. In any case, for most engineering purposes, we are only interested in the mean velocities.

It is therefore convenient to express the instantaneous velocity components in the following way:

$$\left. \begin{aligned} u &= \bar{u} + u', \\ v &= \bar{v} + v', \\ w &= \bar{w} + w', \end{aligned} \right\} \tag{14.1.1}$$

where $\bar{u}$, $\bar{v}$, $\bar{w}$ are the mean velocity components (averaged over a short interval of time at a particular point in space) and u', v', w' are the turbulent fluctuating components.

The Navier-Stokes equation for motion in the x direction may be written

$$\rho \frac{\partial u}{\partial t} + \rho u \frac{\partial u}{\partial x} + \rho v \frac{\partial u}{\partial y} + \rho w \frac{\partial u}{\partial z} = \frac{\partial p_{xx}}{\partial x} + \frac{\partial \tau_{yx}}{\partial y} + \frac{\partial \tau_{zx}}{\partial z}, \quad (14.1.2)$$

and if the equation of continuity for incompressible flow is multiplied by ρu, we can say

$$\rho u \left(\frac{\partial u}{\partial x} + \frac{\partial v}{\partial y} + \frac{\partial w}{\partial z} \right) = 0, \tag{14.1.3}$$

and hence, adding (14.1.2) and (14.1.3), we get

$$\rho \frac{\partial u}{\partial t} + \rho \frac{\partial (u^2)}{\partial x} + \rho \frac{\partial (uv)}{\partial y} + \rho \frac{\partial (uw)}{\partial z} = \frac{\partial p_{xx}}{\partial x} + \frac{\partial \tau_{yx}}{\partial y} + \frac{\partial \tau_{zx}}{\partial z}$$

or
$$\rho \frac{\partial u}{\partial t} = \frac{\partial}{\partial x} (p_{xx} - \rho u^2) + \frac{\partial}{\partial y} (\tau_{yx} - \rho uv) + \frac{\partial}{\partial z} (\tau_{zx} - \rho uw).$$

$$\tag{14.1.4}$$

This is simply another form for the equation of motion in the x direction, and must always be valid for instantaneous velocities.

Similar equations may also be written down for motion in the y and z directions.

If we now substitute in (14.1.4) for the instantaneous velocity components from (14.1.1) we will have

$$\rho \frac{\partial}{\partial t}(\bar{u} + u') = \frac{\partial}{\partial x}(p_{xx} - \rho\bar{u}^2 - \rho u'^2 - 2\rho\bar{u}u')$$

$$+ \frac{\partial}{\partial y}(\tau_{yx} - \rho\bar{u}\bar{v} - \rho u'v' - \rho\bar{u}v' - \rho u'\bar{v})$$

$$+ \frac{\partial}{\partial z}(\tau_{zx} - \rho\bar{u}\bar{w} - \rho u'w' - \rho\bar{u}w' - \rho u'\bar{w}),$$

and if every term in the equation is averaged over a short interval of time, noting that $\overline{u'} = 0$, $\overline{v'} = 0$ and $\overline{w'} = 0$,

$$\rho \frac{\partial \bar{u}}{\partial t} = \frac{\partial}{\partial x}(p_{xx} - \rho\bar{u}^2 - \rho\overline{u'^2}) + \frac{\partial}{\partial y}(\tau_{yx} - \rho\bar{u}\bar{v} - \rho\overline{u'v'})$$

$$+ \frac{\partial}{\partial z}(\tau_{zx} - \rho\bar{u}\bar{w} - \rho\overline{u'w'}). \qquad (14.1.5)$$

If this equation is compared with (14.1.4) it will be observed that the ordinary equations of motion are still valid with mean velocity components in place of instantaneous velocities provided additional stresses are introduced as follows:

$$(p_{xx} - \rho\overline{u'^2}) \quad \text{in place of} \quad p_{xx},$$

$$(\tau_{yx} - \rho\overline{u'v'}) \quad \text{in place of} \quad \tau_{yx},$$

$$(\tau_{zx} - \rho\overline{u'w'}) \quad \text{in place of} \quad \tau_{zx}.$$

$-\rho\overline{u'^2}$, $-\rho\overline{u'v'}$, $-\rho\overline{u'w'}$ are the *Reynolds stresses* and they express mathematically the physical fact that momentum is being transferred by the movement of the fluid associated with fluctuating velocity components u', v', w', in addition to the ordinary transfer of momentum by molecular motion which is normally expressed in terms of viscosity by means of the viscous stresses p_{xx}, τ_{xy}, etc.

The equation of continuity, when averaged over a short interval of time, becomes

$$\frac{\partial \bar{u}}{\partial x} + \frac{\partial \bar{v}}{\partial y} + \frac{\partial \bar{w}}{\partial z} = 0, \qquad (14.1.6)$$

and is therefore unchanged with mean velocities in place of instantaneous values.

An engineer would expect that, just as the theory of viscous motion predicts the relationship between the viscous stresses p_{xx},

τ_{yx}, τ_{zx}, etc., and the components of the rate of strain, any theory of turbulence should be able to express the Reynolds stresses in terms of the mean velocity distribution and its derivatives. Unfortunately, however, no theory of turbulence has yet been evolved which is able to do this.

In view of the apparently random nature of the turbulent fluctuations, it is reasonable to consider a statistical approach to the problem of turbulence. It is possible, for instance, to investigate the correlation between the fluctuating velocity components u' and v' at a particular point. A *correlation coefficient* between u and v can be defined by

$$R = \frac{\overline{u'v'}}{\sqrt{(\overline{u'^2})}\sqrt{(\overline{v'^2})}}. \qquad (14.1.7)$$

With no correlation between u' and v', $\overline{u'v'} = 0$ and $R = 0$. With complete correlation $\overline{u'v'} = \overline{u'^2} = \overline{v'^2}$ so that $R = 1$. In a similar manner a correlation coefficient may be defined for the purpose of relating a particular turbulent velocity component u' at time t with the same velocity component u'_ξ at time $t + \xi$ by

$$R_\xi = \frac{\overline{u'u'_\xi}}{u'^2}, \qquad (14.1.8)$$

and for the correlation between the turbulent velocity component u' and the corresponding component u'_y measured simultaneously at another point in space at distance y, we can say

$$R_y = \frac{\overline{u'u'_y}}{u'^2}. \qquad (14.1.9)$$

These concepts form the starting point for the *statistical theory of turbulence*.

14.2. Mixing length theories

While the statistical approach to turbulence is undoubtedly the correct one, the theory has not yet reached the stage where useful results can be derived that are of direct application to problems of engineering interest. An alternative approach must therefore be used to facilitate the analysis of turbulent flow. The *mixing length theory* is inspired by the kinetic theory of gases and postulates a mechanism for turbulence based on the movement of small discrete lumps or particles of fluid.

Let us suppose that a large number of these hypothetical lumps of fluid are moving backwards and forwards across the stream and carrying with them some *transferable property* M, which may be momentum, heat or mass. We will suppose further that the property M is conserved during the movement of the individual lumps of fluid. In other words, a particle of fluid sets off on its journey and retains its own identity and property until at the end of its travel it suddenly mixes with its new surroundings. It requires a considerable stretch of the imagination to picture a fluid behaving in this way, and it would be much more plausible to suppose that mixing proceeds by continuous movement of the fluid. However, the imaginary mechanism of discontinuous movements is amenable to simple analysis on the lines of the kinetic theory of gases, and the justification for the theory is that it leads to results that can be related to experimental observations. Even if the mechanism is wrong, the results are intelligible.

Consider the case of boundary-layer flow or fully developed flow near the wall of a pipe with mean values for velocity and temperature, etc., which vary only with the distance y measured from the wall, i.e.

$$\bar{M} = f(y). \tag{14.2.1}$$

Suppose a particle of fluid of mass m moves from level y_1 to level y_2 and mixes with the surrounding fluid at the end of its journey. At the same time, to satisfy the equation of continuity, we must suppose that another particle of mass m moves from level y_2 to y_1. If we identify the velocities of these particles with the transverse turbulent velocity component v', we can say that for a large number of such movements the average rate of transfer of the property $\bar{M}$ will be given by

$$\text{rate of transfer} = -\overline{v'(y_2 - y_1)}\frac{\mathrm{d}\bar{M}}{\mathrm{d}y}. \tag{14.2.2}$$
$$\text{(per unit area)}$$

If we define a length l' such that $l'\tilde{v}' = \overline{v'(y_2 - y_1)}$, where $\tilde{v}'$ is the root-mean-square value for the turbulent velocity component, we can rewrite (14.2.2) in the form

$$\text{rate of transfer} = -l'\tilde{v}'\frac{\mathrm{d}\bar{M}}{\mathrm{d}y}. \tag{14.2.3}$$
$$\text{(per unit area)}$$

For the case of *momentum transfer* we have the x momentum as the transferable property, i.e. $\bar{M} = \rho\bar{u}$, and equation (14.2.3) becomes

$$-\tau = -l'\tilde{v}'\frac{\mathrm{d}\bar{M}}{\mathrm{d}y} = -\rho l'v'\frac{\mathrm{d}\bar{u}}{\mathrm{d}y}, \tag{14.2.4}$$

where τ is the turbulent shearing stress.

Note that $l'\tilde{v}'$ has dimensions L^2/T and is sometimes called the *kinematic eddy viscosity* ε, i.e.

$$\frac{\tau}{\rho} = \varepsilon \frac{d\bar{u}}{dy}. \tag{14.2.5}$$

These results should be compared with the corresponding expressions for momentum transfer by molecular motion. According to the kinetic theory of gases, corresponding to (14.2.4), we have

$$\text{rate of transfer} \sim \rho \lambda' \tilde{c} \frac{d\bar{u}}{dy},$$

where λ' is the mean free path, and $\tilde{c}$ is the root-mean-square molecular velocity. Also, corresponding to (14.2.5), we have

$$\frac{\tau}{\rho} = \nu \frac{d\bar{u}}{dy},$$

where ν is the true kinematic viscosity which is proportional to $\lambda'\tilde{c}$ on the kinetic theory.

For the case of *heat transfer* we have $\bar{M} = \rho c_p T$ and for the heat flux q from (14.2.3)

$$q = - l'\tilde{v}' \frac{d}{dy}(\rho c_p T) = - \rho c_p l'\tilde{v}' \frac{dT}{dy}. \tag{14.2.6}$$

For the case of *mass transfer*, if the molal concentration of a diffusing component (A) is C_A, and if the molecular weight of the component is m_A, we can say that the transferable property $\bar{M}$ is the mass of the diffusing component per unit volume of fluid, i.e. $\bar{M} = m_A C_A$, and equation (14.2.3) becomes

$$\text{rate of mass transfer of } (A) \ G_A = - m_A l'\tilde{v}' \frac{dC_A}{dy}, \tag{14.2.7}$$

or, alternatively, the molal transfer rate per unit area, N_A is given by

$$N_A = - l'\tilde{v}' \frac{dC_A}{dy}. \tag{14.2.8}$$

It will be noted that the transfer mechanism outlined above is the same as that postulated in Chapter 11 for the purpose of deriving the Reynolds analogy. From (14.2.4), (14.2.6) and (14.2.7), the analogy may now be extended to include the relationship between skin friction and both heat and mass transfer in fully turbulent flow, i.e.

$$\frac{\tau}{\rho \, d\bar{u}/dy} = - \frac{q/\rho c_p}{dT/dy} = \frac{- G_A}{m_A \, dC_A/dy}. \tag{14.2.9}$$

We require further information about the kinematic eddy viscosity $\overline{l'v'}$, however, before we can carry out any direct calculations on the distribution of velocity, temperature or concentration in turbulent flow.

14.3. The Prandtl momentum transfer theory

Consider again the case of turbulent flow in the vicinity of a plane wall, as indicated in fig. 83. Prandtl's momentum transfer theory is based on the assumption that momentum is conserved in the transfer process. This implies that a lump of fluid moving from level y_1 to y_2 carries with it the mean x momentum appropriate to level y_1. The

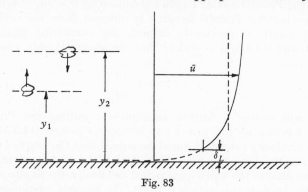

Fig. 83

instantaneous excess of x momentum at the level y_2 due to the turbulent motion is therefore given by

$$\rho u' = - (y_2 - y_1) \frac{\mathrm{d}}{\mathrm{d}y} (\rho \bar{u}). \qquad (14.3.1)$$

If this expression is multiplied by the instantaneous fluctuating velocity component v', and if a mean value is then taken over a short interval of time, we arrive at the Reynolds stress or rate of transfer of momentum across unit area by turbulence, i.e.

$$\tau = - \rho \overline{u'v'} = \overline{v'(y_2 - y_1)} \frac{\mathrm{d}}{\mathrm{d}y} (\rho \bar{u}), \qquad (14.3.2)$$

and this is identical with the general equation (14.2.2) for rate of transfer according to the mixing length theory.

In Prandtl's theory, however, the additional assumption is made that there is complete correlation between the fluctuating velocity components u' and v', i.e. that

$$- \overline{u'v'} = \overline{u'^2} = \overline{v'^2}. \qquad (14.3.3)$$

Then from (14.3.1),

$$u' = -(y_2 - y_1)\frac{d\bar{u}}{dy},$$

and hence,

$$-\overline{u'v'} = \overline{u'^2} = \overline{(y_2 - y_1)^2}\left(\frac{d\bar{u}}{dy}\right)^2,$$

i.e.

$$\tau = -\rho\overline{u'v'} = \rho\overline{(y_2 - y_1)^2}\left(\frac{d\bar{u}}{dy}\right)^2 \qquad (14.3.4)$$

or

$$\tau = \rho l^2\left(\frac{d\bar{u}}{dy}\right)^2, \qquad (14.3.5)$$

where the Prandtl mixing length l is defined by $l^2 = \overline{(y_2 - y_1)^2}$.

Note that the Prandtl length l is different from the ordinary mixing length l' previously defined. By comparing equations (14.2.4) and (14.3.5) it is evident that the kinematic eddy viscosity is given by

$$\varepsilon = l'\tilde{v}' = l^2\frac{d\bar{u}}{dy}. \qquad (14.3.6)$$

We still require a further assumption regarding the Prandtl length l before we can proceed to integrate equation (14.3.5) and obtain a velocity profile. It might be argued that the length l must depend on the distance y from the wall, since the mixing movements must decrease as the wall is approached and finally die out altogether when the laminar sub-layer is reached. The simplest relationship in this case would be one of direct proportionality, i.e.

$$l = ky. \qquad (14.3.7)$$

It might alternatively be argued, however, that the length should depend only on the mean velocity profile in the immediate neighbourhood of the point considered. In this case the simplest relationship, which is dimensionally correct, would be

$$l = -k\frac{d\bar{u}/dy}{d^2\bar{u}/dy^2}, \qquad (14.3.8)$$

which is known as Kármán's hypothesis for the Prandtl mixing length. In either case k is assumed to be a universal constant.

We can now calculate a velocity distribution for *flow near a plane wall* using the Prandtl momentum transfer theory. Strictly speaking, for *boundary-layer flow* with the pressure gradient $\partial p/\partial x = 0$, the equation of motion to be solved is

$$\bar{u}\frac{\partial\bar{u}}{\partial x} + \bar{v}\frac{\partial\bar{u}}{\partial y} = \frac{1}{\rho}\frac{\partial\tau}{\partial y}. \qquad (14.3.9)$$

However, we can ignore the term $\partial \bar{u}/\partial x$ and we can argue that the mean transverse velocity component $\bar{v} = 0$ (although this is strictly true only for fully developed flow in a pipe and not for the case of a boundary layer whose thickness is increasing as the flow proceeds in the x direction). With these simplifying assumptions equation (14.3.9) reduces to the statement

$$\tau = \text{constant} = \tau_0,$$

and hence, from (14.3.5)

$$l^2 \left(\frac{d\bar{u}}{dy}\right)^2 = \frac{\tau_0}{\rho}$$

or

$$l \frac{d\bar{u}}{dy} = \sqrt{\frac{\tau_0}{\rho}}, \tag{14.3.10}$$

and substituting for the length l from (14.3.7) we have

$$ky \frac{d\bar{u}}{dy} = \sqrt{\frac{\tau_0}{\rho}};$$

hence, integrating

$$\bar{u} = \frac{1}{k} \sqrt{\frac{\tau_0}{\rho}} \log_e y + \text{constant}, \tag{14.3.11}$$

and if we approach from the outside of the boundary layer with $\bar{u} = \bar{u}_1$ at $y = \delta$, the equation becomes

$$\bar{u}_1 - \bar{u} = \frac{1}{k} \sqrt{\frac{\tau_0}{\rho}} \log_e \frac{\delta}{y}, \tag{14.3.12}$$

which is a special case of the *velocity defect law*.

Note that with this profile as $y \to 0$, $d\bar{u}/dy \to \infty$ and $\bar{u} \to -\infty$ as indicated by the dotted continuation of the curve in fig. 83. This apparent objection, however, is not serious because the profile of (14.3.12) only refers to the turbulent region *outside the laminar sub-layer*.

We can arrive at the same velocity defect law using Kármán's hypothesis for the mixing length in the Prandtl momentum transfer theory. From (14.3.8) and (14.3.10)

$$k \frac{(d\bar{u}/dy)^2}{d^2\bar{u}/dy^2} = -\sqrt{\frac{\tau_0}{\rho}}$$

or

$$\frac{d^2\bar{u}/dy^2}{(d\bar{u}/dy)^2} = -\frac{k}{\sqrt{(\tau_0/\rho)}};$$

hence, integrating

$$\frac{1}{\mathrm{d}\bar{u}/\mathrm{d}y} = \frac{ky}{\sqrt{(\tau_0/\rho)}} + \text{constant},$$

and the constant may be put equal to zero, implying that if the turbulent velocity profile was extended right to the wall the gradient would be infinite at $y = 0$. Hence

$$\frac{\mathrm{d}\bar{u}}{\mathrm{d}y} = \frac{1}{ky}\sqrt{(\tau_0/\rho)},$$

and, integrating from the outside of the boundary layer with $u = u_1$ at $y = \delta$, we get

$$\bar{u}_1 - \bar{u} = \frac{1}{k}\sqrt{\frac{\tau_0}{\rho}}\log_e\frac{\delta}{y},$$

which is identical with (14.3.12).

The velocity defect law (14.3.12) is confirmed experimentally, not only for flow in the outer portion of a turbulent boundary layer, but also for the central region of the flow in a circular pipe with the pipe radius taking the place of the boundary-layer thickness, i.e.

$$\bar{u}_1 - \bar{u} = \frac{1}{k}\sqrt{\frac{\tau_0}{\rho}}\log_e\frac{a}{y}. \tag{14.3.13}$$

Experimental measurement of the velocity distribution in either case suggests a numerical value for the constant k in the neighbourhood of 0·40.

14.4. The universal velocity profile

We now launch an attack from a different quarter by investigating the turbulent velocity profile in the immediate neighbourhood of the laminar sub-layer.

It could be argued on dimensional grounds that for flow at the inner part of the turbulent region, close to the edge of the laminar sub-layer, the mean velocity $\bar{u}$ at distance y from the wall should be some function of ρ, μ, τ_0 and y, and by the normal methods of dimensional analysis this functional relationship may be reduced to the statement

$$\frac{\bar{u}}{\sqrt{(\tau_0/\rho)}} = f\left(\frac{y\sqrt{(\tau_0/\rho)}}{\nu}\right). \tag{14.4.1}$$

This equation and the velocity defect law in the form of (14.3.11) may be combined to give

$$\frac{\bar{u}}{\sqrt{(\tau_0/\rho)}} = A + \frac{1}{k}\log_e\left[\frac{y\sqrt{(\tau_0/\rho)}}{\nu}\right]. \tag{14.4.2}$$

Writing $u^* = \dfrac{\bar{u}}{\sqrt{(\tau_0/\rho)}}$ and $y^* = \dfrac{y\sqrt{(\tau_0/\rho)}}{\nu}$, equation (14.4.2) becomes

$$u^* = A + B \log_e y^*, \tag{14.4.3}$$

where A and B are constants. Equation (14.4.3) is the so-called *universal velocity profile*. Note that $B = 1/k$, where k is the constant in the expression for the Prandtl mixing length. Note also that, if c_f is defined by $c_f = \tau_0/\frac{1}{2}\rho u^2$,

$$u^* = \frac{\bar{u}}{\sqrt{(\tau_0/\rho)}} = \frac{\bar{u}}{u_1} \sqrt{\frac{2}{c_f}},$$

and

$$y^* = \frac{y\sqrt{(\tau_0/\rho)}}{\nu} = \frac{u_1 y}{\nu} \sqrt{\frac{c_f}{2}}.$$

Good agreement is obtained with experimental measurements of turbulent velocity profiles if we take $A = 5{\cdot}5$ and $B = 2{\cdot}5$ in equation (14.4.3). The value of $2{\cdot}5$ for B corresponds to $k = 0{\cdot}40$, i.e.

$$u^* = 5{\cdot}5 + 2{\cdot}5 \log_e y^* \tag{14.4.3a}$$

or

$$u^* = 5{\cdot}5 + 5{\cdot}75 \log_{10} y^*. \tag{14.4.3b}$$

In the *laminar sub-layer*, if $\tau = \text{constant} = \tau_0$, we must have a linear velocity distribution given by $\tau_0 = \mu u/y$

$$u = \frac{y}{\nu} \frac{\tau_0}{\rho}$$

or

$$\frac{u}{\sqrt{(\tau_0/\rho)}} = \frac{y\sqrt{(\tau_0/\rho)}}{\nu},$$

i.e.

$$u^* = y^*. \tag{14.4.4}$$

The two profiles intersect at $y^* = 11{\cdot}6$ as shown on fig. 84. This gives an indication of the thickness of the laminar sub-layer which is in close agreement with the rough estimate obtained in §13.6.

Although the agreement between the universal turbulent profile (14.4.3) and experimental observations is very good indeed for values of y^* greater than about 30, there is a progressive divergence for values of y^* less than 30. Nikuradse, Kármán and others have suggested that the flow should be divided into three zones as follows:

$y^* < 5$, the true laminar sub-layer with $u^* = y^*$.

$5 < y^* < 30$, a buffer zone for which Kármán suggests that $u^* = -3{\cdot}05 + 5{\cdot}0 \log_e y^*$.

$y^* > 30$, the turbulent zone in which the universal velocity profile (14.4.3) applies.

Doubts have been expressed regarding the validity of the original experimental observations which gave the figure of $y^* = 5$ for the thickness of the true laminar sub-layer. Other indirect measurements, however, have tended to support this figure. It has also been suggested that the laminar sub-layer is not in fact of constant thickness, but that it oscillates in an irregular manner. The main

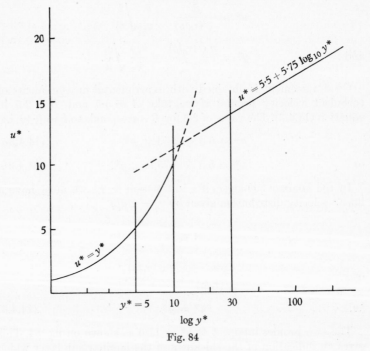

Fig. 84

significance of the value $y^* = 5$ is that it gives an indication of the effect of *surface roughness* on the flow. Projections on the surface having dimensions which correspond to values of y^* less than 5 will be entirely submerged within the laminar sub-layer. The expression 'aerodynamically smooth' is used to describe a surface which satisfies this condition.

If, on the other hand, projections on the surface extend beyond $y^* = 30$ they will begin to penetrate like mountain peaks into the upper turbulent zone and will contribute some additional *form drag* to the ordinary surface friction. Surfaces with projections extending to $y^* = 100$ and above are usually regarded as fully roughened.

14.5. Logarithmic resistance formulae for turbulent flow in pipes

For flow in the turbulent core of a long pipe, it might be argued that the velocity defect $(\bar{u}_1 - \bar{u})$ should be a function of ρ, τ_0, a and y, and hence by the methods of dimensional analysis

$$\frac{\bar{u}_1 - \bar{u}}{\sqrt{(\tau_0/\rho)}} = f\left(\frac{y}{a}\right). \tag{14.5.1}$$

This is simply a generalization of the velocity defect law of (14.3.12) or (14.3.13) which was derived from the Prandtl momentum transfer theory.

If equation (14.5.1) is assumed to apply with reasonable accuracy over the entire cross-section of the flow, it may be integrated to give a spatial mean value u_m for the mean velocity taken over the whole stream, with the result that

$$\frac{\bar{u}_1 - \bar{u}_m}{\sqrt{(\tau_0/\rho)}} = \text{constant.} \tag{14.5.2}$$

It should be noted, however, that the velocity defect law (14.5.1) does not strictly apply in the boundary-layer region close to the wall. The logarithmic form (14.3.13) of the velocity defect law cannot in any case be integrated right up to the wall because of the difficulty which has already been noted that $u \to -\infty$ as $y \to 0$. Nevertheless, the conclusion of (14.5.2) can be accepted even though the numerical value of the constant cannot be calculated directly from the particular form of equation (14.5.1).

For the case of flow in a smooth pipe we can substitute for u_1 from the universal velocity profile of (14.4.2) by putting y equal to a, i.e.

$$\frac{\bar{u}_1}{\sqrt{(\tau_0/\rho)}} = A + \frac{1}{k} \log_e \left(\frac{a\sqrt{(\tau_0/\rho)}}{\nu}\right), \tag{14.5.3}$$

hence, from (14.5.2) and (14.5.3)

$$\frac{\bar{u}_m}{\sqrt{(\tau_0/\rho)}} = \text{constant} + \frac{1}{k} \log_e \left(\frac{a\sqrt{(\tau_0/\rho)}}{\nu}\right), \tag{14.5.4}$$

and if c_f for a pipe is defined by $c_f = \tau_0/\frac{1}{2}\rho\bar{u}_m^2$,

$$\sqrt{\frac{\tau_0}{\rho}} = \bar{u}_m \sqrt{\frac{c_f}{2}},$$

hence, from (14.5.4)

$$\sqrt{\frac{2}{c_f}} = \text{constant} + \frac{1}{k} \log_e \left(\frac{a \bar{u}_m}{\nu} \sqrt{\frac{c_f}{2}} \right)$$

or

$$\sqrt{\frac{1}{c_f}} = C + \frac{1}{k \sqrt{2}} \log_e (Re \sqrt{c_f}), \tag{14.5.5}$$

where C is a constant, and Re is the ordinary Reynolds number for flow in a pipe defined by $\bar{u}_m d / \nu$ or $2 \bar{u}_m a / \nu$. Experimentally, according to Kármán and Nikuradse, it is found that

$$\sqrt{\frac{1}{c_f}} = -0.40 + 1.74 \log_e (Re \sqrt{c_f})$$

or

$$\sqrt{\frac{1}{c_f}} = -0.40 + 4.0 \log_{10} (Re \sqrt{c_f}), \tag{14.5.6}$$

and the factor 4.0 corresponds to a value of $k = 0.408$ which is quite close to the value 0.40 previously used in connexion with the universal velocity profile.

14.6. The velocity profile and resistance formula for rough pipes

For flow in the turbulent core of a *rough pipe*, we can again make use of the velocity defect law represented by (14.5.1) and (14.5.2). The universal velocity profile, however, takes a different form in this case. We could argue on dimensional grounds that for flow near the wall the mean velocity $\bar{u}$ at distance y should be some function of ρ, ε, τ_0 and y, and by the normal methods of dimensional analysis this functional relationship may be reduced to the statement

$$\frac{\bar{u}}{\sqrt{(\tau_0/\rho)}} = f \left(\frac{y}{\varepsilon} \right), \tag{14.6.1}$$

where ε is the equivalent size of the surface roughness. Combining (14.6.1) with the Prandtl form of the velocity defect law (14.3.11), we arrive at the universal velocity profile for flow in a rough pipe:

$$\frac{\bar{u}}{\sqrt{(\tau_0/\rho)}} = A' + \frac{1}{k} \log_e \left(\frac{y}{\varepsilon} \right) \tag{14.6.2}$$

which takes the place of (14.4.2). Putting y equal to a in (14.6.2) gives

$$\frac{\bar{u}_1}{\sqrt{(\tau_0/\rho)}} = A' + \frac{1}{k} \log_e \left(\frac{a}{\varepsilon} \right), \tag{14.6.3}$$

and hence, from (14.5.2) and (14.6.3)

$$\frac{\bar{u}_m}{\sqrt{(\tau_0/\rho)}} = \text{constant} + \frac{1}{k} \log_e \left(\frac{a}{\varepsilon}\right), \qquad (14.6.4)$$

and noting that $\sqrt{\dfrac{\tau_0}{\rho}} = \bar{u}_m \sqrt{\dfrac{c_f}{2}}$ the last result may be written

$$\sqrt{\frac{2}{c_f}} = \text{constant} + \frac{1}{k} \log_e \left(\frac{a}{\varepsilon}\right)$$

or

$$\sqrt{\frac{1}{c_f}} = C' + \frac{1}{k\sqrt{2}} \log_e \left(\frac{a}{\varepsilon}\right). \qquad (14.6.5)$$

Experimentally it is found that the constant C' has the value 3·46, and hence, the equation for the friction coefficient may be expressed as

$$\sqrt{\frac{1}{c_f}} = 3\cdot46 + 1\cdot74 \log_e \left(\frac{a}{\varepsilon}\right)$$

or

$$\sqrt{\frac{1}{c_f}} = 3\cdot46 + 4\cdot0 \log_{10} \left(\frac{a}{\varepsilon}\right), \qquad (14.6.6)$$

which is the rough pipe equivalent of equation (14.5.6) for smooth pipes.

REFERENCES

(1) GOLDSTEIN. *Modern Developments in Fluid Dynamics* (Oxford).
(2) PRANDTL. *The Essentials of Fluid Dynamics* (Blackie).
(3) TAYLOR. Statistical theory of turbulence. *Proc. Roy. Soc.* A, **151**.
(4) BATCHELOR. *Homogeneous Turbulence* (Cambridge University Press).

CHAPTER 15

DIFFUSION AND MASS TRANSFER

15.1. The diffusion law

The experimental law governing the diffusion of one component (A) in a stagnant gas or liquid can be expressed approximately by

$$\mathbf{G}_A = - \mathscr{D} \operatorname{grad} c_A, \qquad (15.1.1)$$

where $\mathbf{G}_A$ is the rate of diffusion or mass flow per unit area, $\mathscr{D}$ is the diffusivity, and c_A is the mass concentration of the diffusing component, i.e. the mass of component (A) per unit volume of fluid. Note that (15.1.1) is similar to the heat conduction law (8.1.1) or (8.1.2). The dimensions of the diffusivity $\mathscr{D}$ are $L^2 T^{-1}$.

Equation (15.1.1) may be expressed alternatively in terms of molal quantities

$$\mathbf{N}_A = - \mathscr{D} \operatorname{grad} C_A, \qquad (15.1.2)$$

where $\mathbf{N}_A$ is the molal rate of diffusion of component (A), and C_A is the molal concentration, i.e. mols of (A) per unit volume of fluid. For a perfect gas the partial pressure is related to the molal concentration by

$$p_A = C_A \bar{R} T, \qquad (15.1.3)$$

where $\bar{R}$ is the universal gas constant. For the diffusion of a gas under conditions of constant temperature, therefore, equation (15.1.2) may be expressed as

$$\mathbf{N}_A = - \frac{\mathscr{D}}{\bar{R} T} \operatorname{grad} p_A. \qquad (15.1.4)$$

The simple form of the diffusion law, expressed by (15.1.1), (15.1.2) or (15.1.4), is strictly only valid when the concentration of the diffusing component is small.

15.2. Diffusion of a gas according to the kinetic theory

Consider the case of two gases (A) and (B) diffusing through each other in the direction of the z-axis. Let w_0 be the bulk velocity of the mixed gases in this direction. It is shown in [1] that the rate of molecular flow of (A) per unit area is given by

$$N_A = n_A w_0 - \tfrac{1}{3} \lambda_A \frac{\partial n_A}{\partial z} \bar{v}_A, \qquad (15.2.1)$$

where n_A is the molecular density of the gas (A), λ_A is the mean free path for the molecules of (A), and $\bar{v}_A$ is the mean molecular velocity for the gas (A). Similarly the rate of molecular flow of (B) per unit area is

$$N_B = n_B w_0 - \tfrac{1}{3}\lambda_B \frac{\partial n_B}{\partial z}\,\bar{v}_B. \tag{15.2.2}$$

Under the condition of constant total pressure, the combined molecular density of the gas mixture must be constant, i.e.

$$n_A + n_B = \text{constant} = n$$

or
$$\frac{\partial n_A}{\partial z} + \frac{\partial n_B}{\partial z} = 0. \tag{15.2.3}$$

Hence, eliminating w_0 between (15.2.1) and (15.2.2), and making use of (15.2.3), we have

$$n_B N_A - n_A N_B = -\tfrac{1}{3}(n_B \lambda_A \bar{v}_A + n_A \lambda_B \bar{v}_B)\frac{\partial n_A}{\partial z}. \tag{15.2.4}$$

For the special case of *equal and opposite molal diffusion* of the two gases, we must have $N_A = -N_B$, and therefore

$$N_A = -\frac{1}{3}\frac{(n_B \lambda_A \bar{v}_A + n_A \lambda_B \bar{v}_B)}{(n_A + n_B)}\frac{\partial n_A}{\partial z}. \tag{15.2.5}$$

If the diffusivity or coefficient of diffusion of the two gases is defined by

$$\mathscr{D} = \frac{1}{3}\frac{(n_B \lambda_A \bar{v}_A + n_A \lambda_B \bar{v}_B)}{(n_A + n_B)}, \tag{15.2.6}$$

equation (15.2.5) becomes

$$N_A = -\mathscr{D}\frac{\partial n_A}{\partial z}$$

or, generalizing,

$$\mathbf{N}_A = -\mathscr{D}\,\text{grad }n_A, \tag{15.2.7}$$

and this is identical with (15.1.2), since the molal concentration C_A and the molecular density n_A are the same thing.

For the special case of the *diffusion of one gas (A) through a stagnant gas (B)*, we must have $N_B = 0$ and therefore from (15.2.4)

$$N_A = -\frac{1}{3}\frac{(n_B \lambda_A \bar{v}_A + n_A \lambda_B \bar{v}_B)}{n_B}\frac{\partial n_A}{\partial z}$$

or
$$\mathbf{N}_A = -\frac{n}{n_B}\mathscr{D}\,\text{grad }n_A, \tag{15.2.8}$$

where the diffusivity $\mathscr{D}$ is defined by equation (15.2.6) as before. Equation (15.2.8) may be expressed in the alternative forms

$$\mathbf{N}_A = -\frac{p}{p_B}\mathscr{D}\ \mathrm{grad}\ n_A \qquad (15.2.9)$$

or

$$\mathbf{N}_A = -\frac{p}{p_B}\frac{\mathscr{D}}{\bar{R}T}\ \mathrm{grad}\ p_A, \qquad (15.2.10)$$

where p_B is the partial pressure of the stagnant gas (B) and p is the total pressure of the gas mixture, assumed constant. If the partial pressure p_A of the diffusing component (A) is small, p_B will be nearly equal to p, and in this case equation (15.2.10) can be replaced by (15.1.4), i.e.

$$\mathbf{N}_A = -\frac{\mathscr{D}}{\bar{R}T}\ \mathrm{grad}\ p_A.$$

15.3. Steady molecular diffusion in one dimension

For steady molecular diffusion in one dimension over a finite distance equation (15.1.2) can be integrated directly to give

$$N_A = \frac{\mathscr{D}(C_{A_1} - C_{A_2})}{z}, \qquad (15.3.1)$$

and similarly, from (15.1.4) in terms of partial pressures

$$N_A = \frac{\mathscr{D}}{\bar{R}T}\frac{(p_{A_1} - p_{A_2})}{z}. \qquad (15.3.2)$$

These results are valid approximately for diffusion in liquids and for diffusion in gases respectively, provided the concentration or partial pressure of the diffusing component is small. The expressions are similar to the corresponding one for the steady conduction of heat through a solid in one dimension.

For the case of diffusion through a stagnant gas when the partial pressure p_A of the diffusing component is not negligible, however, it is necessary to use equation (15.2.9) or (15.2.10). The one-dimensional form of (15.2.10) is

$$N_A = -\frac{p}{p_B}\frac{\mathscr{D}}{\bar{R}T}\frac{\mathrm{d}p_A}{\mathrm{d}z}, \qquad (15.3.3)$$

and under conditions of constant total pressure, $p_A + p_B = p$, i.e.

$$\frac{\mathrm{d}p_A}{\mathrm{d}z} = -\frac{\mathrm{d}p_B}{\mathrm{d}z}, \qquad (15.3.4)$$

so that (15.3.3) can be written

$$N_A = + \frac{p}{p_B} \frac{\mathscr{D}}{\bar{R}T} \frac{\mathrm{d}p_B}{\mathrm{d}z}.$$

Integrating over the distance z from p_{B_1} to p_{B_2} gives

$$N_A = \frac{p\mathscr{D}}{\bar{R}Tz} \log_e \frac{p_{B2}}{p_{B1}} \tag{15.3.5}$$

or $\quad N_A = \frac{\mathscr{D}}{\bar{R}T} \frac{p}{p_{Bm}} \frac{(p_{B2} - p_{B1})}{z} = \frac{\mathscr{D}}{\bar{R}T} \frac{p}{p_{Bm}} \frac{(p_{A1} - p_{A2})}{z},$ (15.3.6)

where p_{Bm} is the logarithmic mean partial pressure for the stagnant gas (B).

15.4. The equation of continuity for mass transfer in a fluid

Consider a control surface S enclosing a volume V in a fluid stream, as shown in fig. 85. If mass transfer of a component (A) is taking

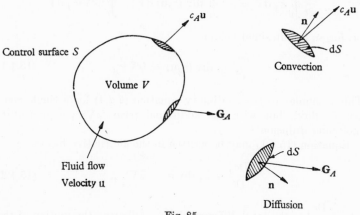

Fig. 85

place by diffusion, there will be a mass flow $\mathbf{G}_A$ per unit area at any point on the control surface which may be expressed in the form of equation (15.1.1), i.e.

$$\mathbf{G}_A = - \mathscr{D} \operatorname{grad} c_A,$$

where c_A is the mass concentration or density of the substance (A). The rate of accumulation of (A) inside the surface S is given by

$$\frac{\partial}{\partial t} \oint_V c_A \, \mathrm{d}V.$$

The net rate of outflow of (A) across the surface by *convection* associated with the bulk velocity $\mathbf{u}$ of the mixed fluid is given by

$$\oint_S c_A \mathbf{u} \cdot \mathrm{d}S = \oint_V \operatorname{div}(c_A\mathbf{u})\,\mathrm{d}V.$$

The net rate of outflow of (A) across the surface by *diffusion* is given by

$$\oint_S \mathbf{G}_A \cdot \mathrm{d}S = \oint_V \operatorname{div}\mathbf{G}_A\,\mathrm{d}V,$$

and substituting for $\mathbf{G}_A$ from (15.1.1) the last expression becomes

$$-\oint_V \mathscr{D}\nabla^2 c_A\,\mathrm{d}V.$$

For conservation of mass of the diffusing component (A), therefore, we must have

$$\frac{\partial}{\partial t}\oint_V c_A\,\mathrm{d}V = -\oint_V \operatorname{div}(c_A\mathbf{u})\,\mathrm{d}V + \oint_V \mathscr{D}\nabla^2 c_A\,\mathrm{d}V$$

or, for an infinitesimal volume,

$$\frac{\partial c_A}{\partial t} + \operatorname{div}(c_A\mathbf{u}) = \mathscr{D}\nabla^2 c_A. \tag{15.4.1}$$

This is similar to the continuity equation (2.2.4) for a single component fluid, but with the additional term $\mathscr{D}\nabla^2 c_A$ representing molecular diffusion.

Equation (15.4.1) may be written in the alternative form†

$$\frac{\mathrm{D}c_A}{\mathrm{D}t} + c_A \operatorname{div}\mathbf{u} = \mathscr{D}\nabla^2 c_A, \tag{15.4.2}$$

where $\dfrac{\mathrm{D}c_A}{\mathrm{D}t}$ is the total differential of c_A following the motion of the fluid, i.e.

$$\frac{\mathrm{D}c_A}{\mathrm{D}t} = \frac{\partial c_A}{\partial t} + \mathbf{u} \cdot \nabla c_A.$$

If the fluid is incompressible, i.e. if the total density of the fluid mixture is uniform, $\operatorname{div}\mathbf{u} = 0$, and equation (15.4.2) becomes

$$\frac{\partial c_A}{\partial t} + \mathbf{u} \cdot \nabla c_A = \mathscr{D}\nabla^2 c_A, \tag{15.4.3}$$

† Noting that $\operatorname{div}(c\mathbf{u}) = \mathbf{u} \cdot \operatorname{grad} c + c \operatorname{div}\mathbf{u}$.

and this may be regarded as the basic mass-transfer equation in fluid flow. With steady conditions the equation becomes

$$\mathbf{u} \cdot \nabla c_A = \mathscr{D} \nabla^2 c_A. \tag{15.4.4}$$

15.5. Similarity for mass transfer in fluid flow

The method employed in §12.4 to investigate the conditions for dynamical similarity between different cases of fluid flow may now be applied to the problem of mass transfer in a fluid. Considering two cases of flow with geometrically similar boundaries, we require to find the conditions for similarity in mass transfer. Let U_1 be a representative velocity, let L be a representative length, and let ρ be the density of the fluid mixture. Introduce dimensionless ratios as follows:

$$u' = \frac{u}{U_1} \text{ etc.,} \quad x' = \frac{x}{L} \text{ etc.,}$$

$$t' = \frac{U_1 t}{L}, \qquad c'_A = \frac{c_A}{\rho}.$$

Equation (15.4.3) may now be expressed in dimensionless form:

$$\frac{\rho U_1}{L} \frac{\partial c'_A}{\partial t'} + \frac{\rho U_1}{L} (\mathbf{u}' \cdot \nabla' c'_A) = \frac{\rho \mathscr{D}}{L^2} \nabla'^2 c$$

or

$$\frac{\partial c'_A}{\partial t'} + \mathbf{u}' \cdot \nabla' c'_A = \frac{\mathscr{D}}{U_1 L} \nabla'^2 c'_A. \tag{15.5.1}$$

For given boundary conditions the solution of this equation, and therefore the mass-transfer pattern, depends only on the ratio $\mathscr{D}/U_1 L$. The inverse of this ratio $U_1 L/\mathscr{D}$ is reminiscent of the Reynolds number $U_1 L/\nu$. It provides a measure of the relative magnitude of mass transfer by convection to mass transfer by diffusion.

Noting that $\dfrac{U_1 L}{\mathscr{D}} = \dfrac{U_1 L}{\nu} \dfrac{\nu}{\mathscr{D}}$ we can say that the condition for complete similarity in fluid flow and mass transfer, between two different cases of forced convection with geometrically similar boundaries, is that the Reynolds numbers should be equal and that the ratio $\nu/\mathscr{D}$ should have the same value in each case. The ratio $\nu/\mathscr{D}$ is known as the *Schmidt number Sc* and is analogous to the Prandtl number in heat transfer. Like the Prandtl number it involves only the physical properties of the fluid.

Referring to the expression (15.1.1) for the diffusion law, a *mass-transfer coefficient* could be defined by $G_A L/\mathscr{D} \Delta c_A$, where Δc_A is the overall difference of concentration in the particular problem considered. In molal quantities the ratio would be $N_A L/\mathscr{D}\Delta C_A$. The dimensionless mass-transfer coefficient defined in this way is analogous to the Nusselt number in heat transfer, but with mass or molal flux taking the place of heat flux and with concentration difference instead of temperature difference. We can conclude from (12.4.2) and (15.5.1) that for *mass transfer by forced convection* in geometrically similar systems

$$\frac{G_A L}{\mathscr{D}\Delta c_A} = \frac{N_A L}{\mathscr{D}\Delta C_A} = f(Re, Sc). \tag{15.5.2}$$

If a mass-transfer factor, analogous to the heat-transfer factor h, is defined by

$$k_c = \frac{G_A}{\Delta c_A} = \frac{N_A}{\Delta C_A}, \tag{15.5.3}$$

the dimensionless ratios for the mass-transfer coefficient in (15.5.2) may be written

$$\frac{k_c L}{\mathscr{D}} = f(Re, Sc). \tag{15.5.4}$$

When dealing with *gaseous substances* it is usual to work in terms of partial pressure rather than concentration, and a mass-transfer factor k_G is therefore defined by

$$k_G = \frac{N_A}{\Delta p_A}, \tag{15.5.5}$$

and the dimensionless mass-transfer coefficient of the Nusselt type may be written

$$\frac{k_G L \bar{R} T}{\mathscr{D}} = f(Re, Sc). \tag{15.5.6}$$

An alternative dimensionless mass-transfer coefficient could be defined by

$$\frac{G_A}{U_1 \Delta c_A} = \frac{N_A}{U_1 \Delta C_A} = \frac{k_c}{U_1}, \tag{15.5.7}$$

and this ratio is analogous to the Stanton number in heat transfer. In a similar manner to equation (15.5.2) or (15.5.4), we can say that for mass transfer by forced convection in geometrically similar systems

$$\frac{k_c}{U_1} = f(Re, Sc). \tag{15.5.8}$$

For gaseous substances, working in terms of partial pressure with the mass-transfer factor k_G, equation (15.5.8) becomes

$$\frac{k_G \bar{R} T}{U_1} = f(Re, Sc). \tag{15.5.9}$$

The physical significance of the Stanton type of mass-transfer coefficient k_c/U_1 is that it represents the ratio of the actual mass-transfer rate by the combined processes of diffusion and convection to the mass flow of component (A) in an undisturbed stream flowing with velocity U_1. The physical meaning of the Nusselt type of coefficient $k_c L/\mathcal{D}$, on the other hand, is that it represents the ratio of the actual mass-transfer rate in the problem considered to a hypothetical rate of mass transfer by a process of pure diffusion with concentration gradient $\Delta C_A/L$.

15.6. Mass transfer in turbulent flow

It has already been noted in Chapter 14 that the Reynolds analogy may be extended to include the relationship between mass transfer

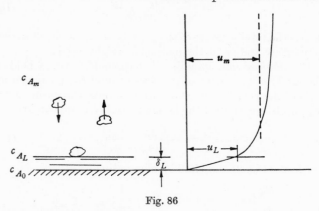

Fig. 86

and skin friction in turbulent flow. From (14.2.9), the relationship may be expressed by

$$\frac{k_c}{u_m} = St = \tfrac{1}{2} c_f. \tag{15.6.1}$$

We can also extend the Taylor-Prandtl analogy in the following way to cover mass transfer. Referring to fig. 86, we will consider the problem of turbulent flow parallel to a plane wall with a laminar sub-layer of thickness δ_L adjacent to the wall.

In the *laminar sub-layer*, assuming a linear concentration and velocity distribution,

$$\tau_0 = \mu \frac{du}{dy} = \mu \frac{u_L}{\delta_L},$$

and

$$G_A = -\mathscr{D} \frac{dc_A}{dy} = -\mathscr{D} \frac{(c_{A_L} - c_{A_0})}{\delta_L},$$

hence,

$$\frac{G_A}{\tau_0} = \frac{-\mathscr{D}(c_{A_L} - c_{A_0})}{\mu u_L} = \frac{(c_{A_0} - c_{A_L})}{\rho u_L Sc}. \tag{15.6.2}$$

For the case of a gas, however, we can use equation (15.3.6) in place of (15.3.1) to derive the diffusion rate through the laminar sub-layer with greater accuracy, i.e.

$$N_A = \frac{\mathscr{D}}{\bar{R}T} \frac{p}{p_{Bm}} \frac{(p_{A_0} - p_{A_L})}{\delta_L},$$

and hence,

$$\frac{G_A}{\tau_0} = \frac{m_A p}{\bar{R}T} \frac{1}{p_{Bm}} \frac{(p_{A_0} - p_{A_L})}{\rho u_L Sc} \tag{15.6.3}$$

or

$$\frac{G_A}{\tau_0} = \frac{p}{p_{Bm}} \frac{(c_{A_0} - c_{A_L})}{\rho u_L Sc}, \tag{15.6.4}$$

where p is the total pressure of the gas mixture, ρ is the mean density of the mixture and m_A is the molecular weight of (A).

In the *turbulent zone*, assuming that Reynolds analogy holds good for the random movement of particles of fluid from a region in the flow where the velocity and concentration have the mean values u_m and c_{A_m} up to the edge of the laminar sub-layer, we can say

$$\frac{G_A}{\tau_0} = -\frac{(c_{A_m} - c_{A_L})}{\rho(u_m - u_L)}. \tag{15.6.5}$$

Hence, from (15.6.2) and (15.6.5), eliminating c_{A_L}

$$\frac{G_A}{\tau_0} [(u_m - u_L) + u_L Sc] = \frac{1}{\rho} (c_{A_0} - c_{A_m}),$$

i.e.

$$\frac{G_A}{u_m \Delta c_A} = \frac{\tau_0}{\rho u_m^2} \frac{1}{1 + \dfrac{u_L}{u_m}(Sc - 1)}$$

or

$$\frac{k_c}{u_m} = \frac{\tfrac{1}{2}c_f}{1 + \dfrac{u_L}{u_m}(Sc - 1)}, \tag{15.6.6}$$

which is the usual form of the Taylor-Prandtl analogy for mass transfer.

For the case of a gas, we can say with slightly greater accuracy, from (15.6.4) and (15.6.5), eliminating c_{A_L},

$$\frac{G_A}{\tau_0}\left[(u_m - u_L) + u_L Sc\frac{p_{Bm}}{p}\right] = \frac{1}{\rho}(c_{A_0} - c_{A_m}),$$

i.e.

$$\frac{G_A}{u_m\Delta c_A} = \frac{k_c}{u_m} = \frac{\frac{1}{2}c_f}{1 + \frac{u_L}{u_m}\left[\frac{p_{Bm}}{p}Sc - 1\right]}$$

or

$$\frac{k_G\bar{R}T}{u_m} = \frac{\frac{1}{2}c_f}{1 + \frac{u_L}{u_m}\left[\frac{p_{Bm}}{p}Sc - 1\right]}. \tag{15.6.7}$$

REFERENCES

(1) JEANS. *Kinetic Theory of Gases* (Cambridge University Press).
(2) SHERWOOD and PIGFORD. *Absorption and Extraction* (McGraw-Hill).

CHAPTER 16

THE ENERGY EQUATION AND HEAT TRANSFER

16.1. The energy equation for an element of fluid

In Chapter 3 the general energy equation was derived for flow along a streamline and for flow in a pipe or duct with fixed walls. In each case, however, an overall result was obtained without going into any details regarding the mechanism of energy transfer. We are now in a position to investigate this mechanism and to derive an energy balance for an individual element of fluid.

Consider a particular element of fluid $dx\,dy\,dz$ surrounded by an imaginary flexible boundary as indicated in fig. 87. The element will be liable to expansion and distortion under the action of the

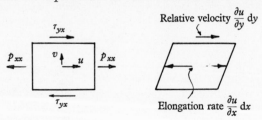

Situation at time t At time $t + dt$

Fig. 87

fluid stresses listed in (12.2.1) and (12.2.2). The total rate of work being done *on* the fluid element by the surface stresses p_{xx}, τ_{yx}, etc., will be the sum of nine terms each involving the product of stress, surface area and relative velocity or rate of strain. Referring to Appendix 4, it will be evident that the total rate of work being done on the element $dx\,dy\,dz$ is given by

$$\sum p_{xx}\,dy\,dz\,\frac{\partial u}{\partial x}\,dx + \sum \tau_{yx}\,dx\,dz\,\frac{\partial u}{\partial y}\,dy$$

or, dividing by $dx\,dy\,dz$, the rate of work *per unit volume* is given by

$$p_{xx}\frac{\partial u}{\partial x} + p_{yy}\frac{\partial v}{\partial y} + p_{zz}\frac{\partial w}{\partial z} + \tau_{xy}\left(\frac{\partial u}{\partial y} + \frac{\partial v}{\partial x}\right)$$
$$+ \tau_{yz}\left(\frac{\partial v}{\partial z} + \frac{\partial w}{\partial y}\right) + \tau_{zx}\left(\frac{\partial w}{\partial x} + \frac{\partial u}{\partial z}\right).$$

198

Substituting for p_{xx} and τ_{xy}, etc., from (12.2.1) and (12.2.2)

rate of work by surface stresses on the fluid per unit volume

$$= - p \operatorname{div} \mathbf{u} + \mu \left[2 \left(\frac{\partial u}{\partial x} \right)^2 + 2 \left(\frac{\partial v}{\partial y} \right)^2 + 2 \left(\frac{\partial w}{\partial z} \right)^2 + \left(\frac{\partial u}{\partial y} + \frac{\partial v}{\partial x} \right)^2 \right.$$

$$\left. + \left(\frac{\partial v}{\partial z} + \frac{\partial w}{\partial y} \right)^2 + \left(\frac{\partial w}{\partial x} + \frac{\partial u}{\partial z} \right)^2 - \frac{2}{3} (\operatorname{div} \mathbf{u})^2 \right]$$

or   rate of work $= - p \operatorname{div} \mathbf{u} + \Phi$,   (16.1.1)

where $\Phi = \mu \left[2 \left(\frac{\partial u}{\partial x} \right)^2 + 2 \left(\frac{\partial v}{\partial y} \right)^2 + 2 \left(\frac{\partial w}{\partial z} \right)^2 + \left(\frac{\partial u}{\partial y} + \frac{\partial v}{\partial x} \right)^2 \right.$

$$\left. + \left(\frac{\partial v}{\partial z} + \frac{\partial w}{\partial y} \right)^2 + \left(\frac{\partial w}{\partial x} + \frac{\partial u}{\partial z} \right)^2 - \frac{2}{3} (\operatorname{div} \mathbf{u})^2 \right],$$

The term $- p \operatorname{div} \mathbf{u}$ in equation (16.1.1) is the rate of work being done by the stresses in *compressing* the fluid. This is the reversible or recoverable work of compression. The term Φ is the rate of work being done by the stresses in *distorting* the fluid, and this work is dissipated within the fluid as 'friction heat'.

We can now follow the particular element of fluid $\mathrm{d}x\,\mathrm{d}y\,\mathrm{d}z$ shown in fig. 87 and apply the first law of thermodynamics from the point of view of an observer who is *outside the element* but moving with it. The only observable heat transfer in this case would be that crossing the boundary of the element by conduction. This is given by

$$- \operatorname{div} \mathbf{q}\, \mathrm{d}x\,\mathrm{d}y\,\mathrm{d}z = Jk\nabla^2 T\, \mathrm{d}x\,\mathrm{d}y\,\mathrm{d}z. \qquad (16.1.2)$$

The first law can then be expressed as follows:

$$\begin{matrix} \text{rate of heat supply} \\ \text{by conduction} \end{matrix} + \begin{matrix} \text{rate of work by} \\ \text{surface stresses} \end{matrix} = \begin{matrix} \text{rate of gain of} \\ \text{internal energy} \end{matrix}$$

$$Jk\nabla^2 T\, \mathrm{d}x\,\mathrm{d}y\,\mathrm{d}z + (- p \operatorname{div} \mathbf{u} + \Phi)\, \mathrm{d}x\,\mathrm{d}y\,\mathrm{d}z = (\rho\, \mathrm{d}x\,\mathrm{d}y\,\mathrm{d}z) \frac{\mathrm{D}E}{\mathrm{D}t},$$

i.e. $Jk\nabla^2 T - p \operatorname{div} \mathbf{u} + \Phi = \rho \dfrac{\mathrm{D}E}{\mathrm{D}t},$

where $\mathrm{D}/\mathrm{D}t$ is the total differential following the motion of the fluid, or

$$\rho \frac{\mathrm{D}E}{\mathrm{D}t} + p \operatorname{div} \mathbf{u} = Jk\nabla^2 T + \Phi. \qquad (16.1.3)$$

The enthalpy H is defined by $H = E + p/\rho$, and hence,

$$\frac{\mathrm{D}H}{\mathrm{D}t} = \frac{\mathrm{D}E}{\mathrm{D}t} + \frac{1}{\rho} \frac{\mathrm{D}p}{\mathrm{D}t} - \frac{p}{\rho^2} \frac{\mathrm{D}\rho}{\mathrm{D}t},$$

but from the continuity equation

$$\frac{D\rho}{Dt} + \rho \operatorname{div} \mathbf{u} = 0,$$

and therefore

$$\rho \frac{DH}{Dt} = \rho \frac{DE}{Dt} + \frac{Dp}{Dt} + p \operatorname{div} \mathbf{u}$$

or

$$\rho \frac{DE}{Dt} + p \operatorname{div} \mathbf{u} = \rho \frac{DH}{Dt} - \frac{Dp}{Dt}. \qquad (16.1.4)$$

Hence, from (16.1.3) and (16.1.4),

$$\rho \frac{DH}{Dt} - \frac{Dp}{Dt} = Jk\nabla^2 T + \Phi, \qquad (16.1.5)$$

which is the general form for the energy equation applied to a particular element of fluid.

In the case of a perfect gas we can substitute $H = Jc_p T$ and equation (16.1.5) becomes

$$\rho Jc_p \frac{DT}{Dt} - \frac{Dp}{Dt} = Jk\nabla^2 T + \Phi. \qquad (16.1.6)$$

If the mechanical energy terms Dp/Dt and Φ in this equation are small, which is the usual state of affairs with heat transfer at low velocities, the energy equation for an element of fluid can be expressed approximately by

$$\frac{DT}{Dt} = \frac{k}{\rho c_p} \nabla^2 T. \qquad (16.1.7)$$

16.2. Similarity for heat transfer in fluid flow

The method of §§12.4 and 15.5 can now be applied to the problem of heat transfer in fluid flow. Considering two cases of flow with geometrically similar boundaries, we require to find the condition for similarity in heat transfer by conduction and forced convection. Let U_1 be a representative velocity, L a representative length, and ΔT a representative temperature difference. Introduce dimensionless ratios as follows:

$$u' = \frac{u}{U_1} \text{ etc.}, \quad x' = \frac{x}{L} \text{ etc.}$$

$$t' = \frac{U_1 t}{L}, \quad p' = \frac{p}{\rho U_1^2},$$

$$H' = \frac{H}{Jc_p \Delta T}, \quad T' = \frac{T}{\Delta T}.$$

The energy equation (16.1.5) for a fluid element may now be expressed in dimensionless form

$$\frac{\rho U_1}{L} Jc_p \Delta T \frac{DH'}{Dt} - \frac{\rho U_1^3}{L} \frac{Dp'}{Dt'} = Jk \frac{\Delta T}{L^2} \nabla'^2 T' + \mu \frac{U_1^2}{L^2} \Phi'$$

or

$$\frac{DH'}{Dt'} - \frac{U_1^2}{Jc_p \Delta T} \frac{Dp'}{Dt'} = \frac{k}{\rho U_1 L c_p} \nabla'^2 T' + \frac{\mu U_1}{\rho c_p J \Delta T L} \Phi', \quad (16.2.1)$$

where

$$\Phi' = \left[2 \left(\frac{\partial u'}{\partial x'} \right)^2 + \text{etc.} \right].$$

Equation (16.2.1) may also be written

$$\frac{DH'}{Dt'} - \frac{U_1^2}{Jc_p \Delta T} \frac{Dp'}{Dt'} = \frac{\alpha}{U_1 L} \nabla'^2 T' + \frac{U_1^2}{Jc_p \Delta T} \frac{\Phi'}{Re}, \quad (16.2.2)$$

where α is the thermal diffusivity defined by $\alpha = k/\rho c_p$. For given boundary conditions the solution of this equation will depend on the ratios $\alpha/U_1 L$ and $U_1^2/Jc_p \Delta T$ in addition to the Reynolds number Re.

The inverse of the first ratio is the *Péclet number* $U_1 L/\alpha$ which is analogous to the ratio $U_1 L/\mathcal{D}$ in mass transfer. It provides a measure of the relative magnitude of heat transfer by convection to heat transfer by conduction. Since the Péclet number $Pe = Re \times Pr$, however, and since the Reynolds number is required independently for dynamical similarity, it is customary to work in terms of the Reynolds and Prandtl numbers rather than the Péclet number. We can therefore deduce from equation (16.2.2) that the condition for complete similarity in fluid flow and heat transfer, between two different cases of forced convection with geometrically similar boundaries, is that the Reynolds number, the Prandtl number, and the ratio $U_1^2/Jc_p \Delta T$ should each have the same values in the two systems.

Taking the Nusselt number hL/k as the dimensionless heat-transfer coefficient, we can therefore say that for forced convection

$$Nu = f \left(Re, Pr, \frac{U_1^2}{Jc_p \Delta T} \right). \quad (16.2.3)$$

Alternatively, using the Stanton number $h/\rho U_1 c_p$,

$$St = f \left(Re, Pr, \frac{U_1^2}{Jc_p \Delta T} \right). \quad (16.2.4)$$

It should be noted, however, that in deriving the dimensionless equation (16.2.2) the fluid properties ρ, μ, k, c_p were taken as being uniform, whereas in fact they are all dependent on the temperature distribution. If temperature differences are large enough to cause

measurable variation in the fluid properties it is necessary to introduce a dimensionless temperature ratio $\Delta T/T_m$ as an additional
criterion in specifying the conditions for similarity.

If the ratio $U_1^2/Jc_p\Delta T$ is small, which is usually the case unless
the flow velocity is very large, equation (16.2.2) will approximate to

$$\frac{DH'}{Dt'} = \frac{\alpha}{U_1 L}\, \nabla'^2\, T',\qquad (16.2.5)$$

and equation (16.2.3) becomes

$$Nu = f(Re,\ Pr).\qquad (16.2.6)$$

Similarly, equation (16.2.4) becomes

$$St = f(Re,\ Pr).\qquad (16.2.7)$$

These are the usual statements regarding similarity in heat transfer
by forced convection with flow at low or moderate velocities. They
were previously derived in Chapter 10 by the method of dimensional
analysis.

16.3. The energy equation for a boundary layer

An energy equation can be derived for boundary-layer flow on the
lines of the momentum equation of §13.4. Considering the case of
flow past a flat surface with heat transfer, we will assume that under

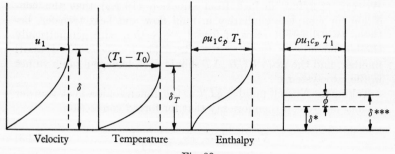

Fig. 88

appropriate conditions the significant changes of temperature in the
fluid are confined to a relatively thin region adjacent to the wall.
We can regard this as a *thermal boundary layer* analogous to the
velocity or momentum boundary layer. Referring to fig. 88, let δ
be the thickness of the velocity layer and let δ_T be that of the
thermal boundary layer. In this particular diagram the thickness δ_T
is shown as being less than δ. This, however, is not necessarily the

case. The enthalpy flow in the velocity boundary layer will be given by

$$H = \int_0^{\delta} \rho u c_p T \, \mathrm{d}y = \rho u_1 c_p T_1 (\delta - \delta^{***}),$$

where δ^{***} is defined by fig. 88, and hence

$$\delta^{***} = \int_0^{\delta} \left(1 - \frac{u}{u_1} \frac{T}{T_1} \right) \mathrm{d}y. \tag{16.3.1}$$

By analogy with the momentum thickness, we can define a quantity ϕ as the *thermal thickness* by $\phi = \delta^{***} - \delta^*$ as shown in fig. 88. Substituting from (16.3.1), the thermal thickness is given by

$$\phi = \int_0^{\delta} \frac{u}{u_1} \left(1 - \frac{T}{T_1} \right) \mathrm{d}y. \tag{16.3.2}$$

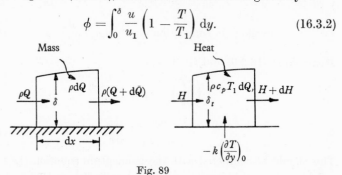

Fig. 89

In fig. 88 the thermal boundary layer is shown as having a smaller thickness than the velocity boundary layer. If this is in fact the case, we can stop the integration short at $y = \delta_T$ in equation (16.3.2), since $T = T_1$ for values of $y > \delta_T$. If, on the other hand, δ_T is greater than δ, we should define H as the enthalpy flow in the thermal boundary layer, i.e.

$$H = \int_0^{\delta_T} \rho u c_p T \, \mathrm{d}y,$$

and this will lead to the definition of the thermal thickness ϕ as

$$\phi = \int_0^{\delta_T} \frac{u}{u_1} \left(1 - \frac{T}{T_1} \right) \mathrm{d}y. \tag{16.3.3}$$

If we now consider a short length $\mathrm{d}x$ of the boundary layer, as shown in fig. 89, we can carry out an energy balance as follows. The mass flow Q and the enthalpy flow H in the boundary layer at distance x from the leading edge are given by

$$\rho Q = \rho u_1 (\delta - \delta^*) \tag{16.3.4}$$

and

$$H = \rho u_1 c_p T_1 (\delta - \delta^* - \phi). \tag{16.3.5}$$

The inflow of heat to the element by conduction across the wall is

$$q_0 \, dx = - k \left(\frac{\partial T}{\partial y} \right)_0 dx. \tag{16.3.6}$$

The net outflow of heat by convection, assuming a constant velocity u_1 and a uniform temperature T_1 outside the boundary layer, is given by

$$\frac{dH}{dx} \, dx - \rho c_p T_1 \frac{dQ}{dx} \, dx = \rho u_1 c_p T_1 \frac{d}{dx} (\delta - \delta^* - \phi) \, dx$$

$$- \rho u_1 c_p T_1 \frac{d}{dx} (\delta - \delta^*) \, dx,$$

i.e.   outflow by convection $= - \rho u_1 c_p T_1 \dfrac{d\phi}{dx} \, dx.$ \tag{16.3.7}

Hence from (16.3.6) and (16.3.7)

$$k \left(\frac{\partial T}{\partial y} \right)_0 = \rho u_1 c_p T_1 \frac{d\phi}{dx}$$

or $- \dfrac{q_0}{\rho u_1 c_p T_1} = \dfrac{d\phi}{dx}.$ \tag{16.3.8}

This should be compared with the momentum equation (13.4.4).

16.4. Approximate solution for heat transfer in a laminar boundary layer

For the velocity profile in a laminar boundary layer with zero pressure gradient we can use the approximate form

$$\frac{u}{u_1} = 2 \left(\frac{y}{\delta} \right) - 2 \left(\frac{y}{\delta} \right)^3 + \left(\frac{y}{\delta} \right)^4,$$

and putting this into the momentum equation we have the result from §13.5 that

$$\delta = 5 \cdot 83 \left(\frac{\nu x}{u_1} \right)^{\frac{1}{2}}.$$

For the thermal boundary layer it will be reasonable to assume a similar form for the temperature distribution

$$\frac{T - T_0}{T_1 - T_0} = 2 \left(\frac{y}{\delta_T} \right) - 2 \left(\frac{y}{\delta_T} \right)^3 + \left(\frac{y}{\delta_T} \right)^4 \tag{16.4.1}$$

or $\dfrac{T_1 - T}{T_1 - T_0} = 1 - 2 \left(\dfrac{y}{\delta_T} \right) + 2 \left(\dfrac{y}{\delta_T} \right)^3 - \left(\dfrac{y}{\delta_T} \right)^4.$

The thermal thickness ϕ will then be given by

$$\frac{\phi}{\delta_t} = \int_0^1 \frac{u}{u_1}\left(1 - \frac{T}{T_1}\right) d\left(\frac{y}{\delta_T}\right) = \frac{T_1 - T_0}{T_1}[\tfrac{2}{15}\,\xi - \tfrac{3}{140}\,\xi^3 + \tfrac{1}{180}\,\xi^4],$$

where ξ is the ratio δ_T/δ. The heat flux at the wall will be given by

$$q_0 = -k\left(\frac{\partial T}{\partial y}\right)_0 = -\frac{2}{\delta_T}k(T_1 - T_0). \qquad (16.4.2)$$

Hence, substituting for q_0 and ϕ in the energy equation

$$2\,\frac{k}{\delta_T}(T_1 - T_0) = \rho u_1 c_p (T_1 - T_0)[\tfrac{2}{15}\,\xi - \tfrac{3}{140}\,\xi^3 + \tfrac{1}{180}\,\xi^4]\frac{d\delta_T}{dx}$$

or
$$\frac{k}{\rho u_1 c_p} = [\tfrac{2}{15}\,\xi - \tfrac{3}{140}\,\xi^3 + \tfrac{1}{180}\,\xi^4]\frac{\delta_T}{2}\frac{d\delta_T}{dx}.$$

Hence, integrating and assuming that the thermal boundary layer starts from the leading edge $x = 0$,

$$\frac{kx}{\rho u_1 c_p} = \frac{\delta_T^2}{4}[\tfrac{2}{15}\,\xi - \tfrac{3}{140}\,\xi^3 + \tfrac{1}{180}\,\xi^4]. \qquad (16.4.3)$$

But for the velocity layer from the momentum equation

$$\delta = 5{\cdot}83\,(vx/u_1)^{\frac{1}{2}} \quad \text{or} \quad \delta^2 = 34\,\mu x/\rho u_1.$$

Therefore (16.4.3) may be written

$$\frac{k}{\mu c_p} = 8{\cdot}5\,\xi^2[\tfrac{2}{15}\,\xi - \tfrac{3}{140}\,\xi^3 + \tfrac{1}{180}\,\xi^4]$$

or
$$\frac{1}{Pr} = \xi^3(1{\cdot}135 - 0{\cdot}182\,\xi^2 + 0{\cdot}047\,\xi^3), \qquad (16.4.4)$$

i.e. the ratio ξ is a function of the Prandtl number. When $Pr = 1$, $\xi = 1$ and the thermal and velocity boundary layers then have the same thickness. If $Pr > 1$, δ_T/δ will be < 1 and vice versa. A reasonably good approximate solution to (16.4.4) for values of Pr greater than $1{\cdot}0$ is

$$\xi = \frac{1}{Pr^{\frac{1}{3}}}. \qquad (16.4.5)$$

Taking this result, the local heat flux from (16.4.2) is given by

$$q_0 = -2k(T_1 - T_0)\frac{Pr^{\frac{1}{3}}}{\delta} = 2k\Delta T\frac{Pr^{\frac{1}{3}}Re^{\frac{1}{2}}}{5{\cdot}83\,x},$$

and hence for the *local Nusselt number*

$$\frac{q_0 x}{k\Delta T} = 0{\cdot}343\,Re^{\frac{1}{2}}Pr^{\frac{1}{3}}. \qquad (16.4.6)$$

The overall rate of heat transfer is given by

$$Q = \int_0^L q_0\, \mathrm{d}x = 0.343 Pr^{\frac{1}{3}} k \Delta T \left(\frac{u_1}{\nu}\right)^{\frac{1}{2}} \int_0^L x^{-\frac{1}{2}}\, \mathrm{d}x,$$

and hence for the overall or *average Nusselt number*

$$Nu = \frac{Q}{k\Delta T} = 0.686 Re^{\frac{1}{2}} Pr^{\frac{1}{3}}. \qquad (16.4.7)$$

Alternatively, in terms of the Stanton number, for the overall or average coefficient,

$$St = 0.686 Re^{-\frac{1}{2}} Pr^{-\frac{2}{3}}. \qquad (16.4.8)$$

Note that in these expressions the Reynolds number is specified in terms of the distance x or length L measured from the leading edge.

16.5. The energy equation with turbulent flow

With turbulent flow both temperatures and velocities will fluctuate. The instantaneous temperature at a point in the flow may be expressed by

$$T = \bar{T} + T', \qquad (16.5.1)$$

where $\bar{T}$ is the mean value and T' is the fluctuating component. The instantaneous velocity components may be similarly expressed as in (14.1.1).

Taking the energy equation in the form of (16.1.7), i.e.

$$\frac{\partial T}{\partial t} + u\frac{\partial T}{\partial x} + v\frac{\partial T}{\partial y} + w\frac{\partial T}{\partial z} = \frac{k}{\rho c_p}\nabla^2 T,$$

substituting from (14.1.1) and (16.5.1) and averaging each term in the equation over a short interval of time, we have the result

$$\frac{\partial \bar{T}}{\partial t} + \bar{u}\frac{\partial \bar{T}}{\partial x} + \frac{\partial}{\partial x}\overline{(u'T')} + \text{etc.} = \frac{k}{\rho c_p}\nabla^2 \bar{T}$$

or $\quad \dfrac{\mathrm{D}\bar{T}}{\mathrm{D}t} = \dfrac{k}{\rho c_p}\nabla^2 \bar{T} - \left\{\dfrac{\partial}{\partial x}\overline{(u'T')} + \dfrac{\partial}{\partial y}\overline{(v'T')} + \dfrac{\partial}{\partial z}\overline{(w'T')}\right\},$ (16.5.2)

which is the ordinary energy equation written in terms of mean values but with the addition of the extra terms in the bracket on the right-hand side. These additional terms involving the fluctuating components represent the transfer of heat by turbulent convection, and they are analogous to the Reynolds stresses in the equation of motion.

Taking the simple case of flow parallel to a plane wall with mean velocity $\bar{u}$ which is a function of y and with heat transfer in the y direction only, equation (16.5.2) reduces to

$$\frac{\mathrm{D}\bar{T}}{\mathrm{D}t} = \frac{k}{\rho c_p} \frac{\partial^2 \bar{T}}{\partial y^2} - \frac{\partial}{\partial y} \overline{(v'T')}$$

or

$$\frac{\mathrm{D}\bar{T}}{\mathrm{D}t} = \frac{1}{\rho c_p} \frac{\partial}{\partial y} \left\{ k \frac{\partial \bar{T}}{\partial y} - \rho c_p \overline{v'T'} \right\}. \tag{16.5.3}$$

The first term on the right-hand side represents the normal transfer of heat by molecular conduction within the fluid, and the second term represents the process of turbulent or eddy transfer. According to the general mixing length theory the eddy heat-transfer rate may be expressed by means of equation (14.2.6), and hence

$$- q_T = \rho c_p \frac{\partial}{\partial y} \overline{(v'T')} = - \rho c_p l' \tilde{v}' \frac{\mathrm{d}\bar{T}}{\mathrm{d}y},$$

where l' is the mixing length and $\tilde{v}'$ is the root-mean-square value of the transverse turbulent velocity component. The total heat flux q due to the combined effects of molecular conduction and eddy transfer within the fluid at any point is therefore given by

$$q = - \left\{ k \frac{\mathrm{d}\bar{T}}{\mathrm{d}y} + \rho c_p l' \tilde{v}' \frac{\mathrm{d}\bar{T}}{\mathrm{d}y} \right\}$$

or

$$q = - \rho c_p (\alpha + \varepsilon) \frac{\mathrm{d}\bar{T}}{\mathrm{d}y}, \tag{16.5.4}$$

where α is the thermal diffusivity $k/\rho c_p$ and ε is the kinematic eddy viscosity $l'\tilde{v}'$.

CHAPTER 17

FORCED CONVECTION

17.1. Forced convection in laminar flow

It will be convenient at this stage to review the basic equations covering fluid mechanics, mass transfer and heat transfer which have been evolved in the preceding chapters.

The *equation of continuity*, which is based on the physical principle of conservation of total mass, states that

$$\frac{\partial \rho}{\partial t} + \text{div}\,(\rho \mathbf{u}) = 0, \qquad (17.1.1)$$

and this may be written

$$\frac{D\rho}{Dt} + \rho\,\text{div}\,\mathbf{u} = 0, \qquad (17.1.1a)$$

and for incompressible flow with ρ constant

$$\text{div}\,\mathbf{u} = 0. \qquad (17.1.1b)$$

The *equation of fluid motion*, based on Newton's laws of motion, together with the experimental relationship between viscous stresses and rates of strain, is

$$\rho\,\frac{D\mathbf{u}}{Dt} = \rho\mathbf{F} - \text{grad}\,p + \mu\nabla^2\mathbf{u} + \frac{\mu}{3}\,\text{grad div}\,\mathbf{u}, \qquad (17.1.2)$$

or for incompressible flow with no free surface

$$\rho\,\frac{D\mathbf{u}}{Dt} = -\,\text{grad}\,p + \mu\nabla^2\mathbf{u}, \qquad (17.1.2a)$$

and for the special case of uniform pressure

$$\frac{D\mathbf{u}}{Dt} = \nu\nabla^2\mathbf{u}. \qquad (17.1.2b)$$

The *mass-transfer equation*, which is based on the principle of the conservation of mass of one component together with the experimental diffusion law, states that

$$\frac{Dc_A}{Dt} + c_A\,\text{div}\,\mathbf{u} = \mathscr{D}\nabla^2 c_A, \qquad (17.1.3)$$

and for uniform total density with div $\mathbf{u} = 0$

$$\frac{\mathrm{D}c_A}{\mathrm{D}t} = \mathscr{D}\nabla^2 c_A. \qquad (17.1.3a)$$

The *energy equation*, which is based on the first law of thermo-dynamics together with the experimental heat conduction law, states that

$$\rho\frac{\mathrm{D}H}{\mathrm{D}t} - \frac{\mathrm{D}p}{\mathrm{D}t} = Jk\nabla^2 T + \Phi \qquad (17.1.4)$$

or, for a perfect gas with $H = Jc_p T$, and neglecting the dissipation function Φ and the term $\mathrm{D}p/\mathrm{D}t$,

$$\frac{\mathrm{D}T}{\mathrm{D}t} = \alpha\nabla^2 T, \qquad (17.1.4a)$$

where α is the thermal diffusivity $k/\rho c_p$.

The resemblance between equations (17.1.2b), (17.1.3a) and (17.1.4a) will be noted. In the special case when $\nu = \mathscr{D} = \alpha$, i.e. if $Pr = Sc = 1$, the three equations will have similar solutions for any particular set of boundary conditions.

17.2. Forced convection in a laminar boundary layer

Consider the case of flow past a flat surface with a laminar boundary layer and zero pressure gradient. Using the method of §13.1 we can examine the orders of magnitude of the various terms in the equations on the assumption that the boundary-layer thickness is small. The four basic equations of continuity, fluid motion, mass transfer and heat transfer, can then be reduced to

$$\frac{\partial u}{\partial x} + \frac{\partial v}{\partial y} = 0, \qquad (17.2.1)$$

$$u\frac{\partial u}{\partial x} + v\frac{\partial u}{\partial y} = \nu\frac{\partial^2 u}{\partial y^2}, \qquad (17.2.2)$$

$$u\frac{\partial c_A}{\partial x} + v\frac{\partial c_A}{\partial y} = \mathscr{D}\frac{\partial^2 c_A}{\partial y^2}, \qquad (17.2.3)$$

$$u\frac{\partial T}{\partial x} + v\frac{\partial T}{\partial y} = \alpha\frac{\partial^2 T}{\partial y^2}. \qquad (17.2.4)$$

The exact solution for the equation of motion (17.2.2) is given in Appendix 5. By introducing the new variables η and f, where $\eta = \frac{1}{2}(u_1/\nu x)^{\frac{1}{2}}y$ and $f = -\psi/(u_1\nu x)^{\frac{1}{2}}$ and ψ is the stream function defined by $u = -\partial\psi/\partial y$ and $v = \partial\psi/\partial x$, the equation of motion (17.2.2) is reduced to the ordinary differential equation

$$\frac{d^3 f}{d\eta^3} + f\frac{d^2 f}{d\eta^2} = 0, \qquad (17.2.5)$$

which can be solved numerically. This equation may also be written

$$\frac{d^2 u}{d\eta^2} + f\frac{du}{d\eta} = 0, \qquad (17.2.6)$$

since $u/u_1 = \frac{1}{2}(df/d\eta)$.

If we make the same substitutions for the velocity components u and v and for the coordinates x and y in equations (17.2.3) and (17.2.4), the mass-transfer and energy equations become

$$\frac{d^2 c_A}{d\eta^2} + \frac{\nu}{\mathscr{D}}f\frac{dc_A}{d\eta} = 0, \qquad (17.2.7)$$

and

$$\frac{d^2 T}{d\eta^2} + \frac{\nu}{\alpha}f\frac{dT}{d\eta} = 0. \qquad (17.2.8)$$

For the special case when $Pr = 1$ and $Sc = 1$, (17.2.7) and (17.2.8) are similar to equation (17.2.6) for the velocity distribution. With the same boundary conditions, therefore, the temperature and concentration profiles must be similar to the velocity profile for the boundary layer.

For momentum, mass and heat transfer at the wall we have

$$\tau_0 = \mu\left(\frac{\partial u}{\partial y}\right)_0, \quad G_A = -\mathscr{D}\left(\frac{\partial c_A}{\partial y}\right)_0 \quad \text{and} \quad q = -k\left(\frac{\partial T}{\partial y}\right)_0.$$

It will therefore follow from the exact solution for the skin friction coefficient derived in Appendix 5 that, with $Pr = Sc = 1$,

$$\frac{\tau_0}{\rho u_1^2} = \frac{G_A}{u_1\Delta c_A} = \frac{q}{\rho u_1 c_p \Delta T} = 0.332\left(\frac{u_1 x}{\nu}\right)^{-\frac{1}{2}}, \qquad (17.2.9)$$

i.e. for mass transfer

$$\frac{G_A}{u_1\Delta c_A} = \tfrac{1}{2}c_f = 0.332 Re^{-\frac{1}{2}},$$

and for heat transfer

$$St = \tfrac{1}{2}c_f = 0.332 Re^{-\frac{1}{2}}.$$

If the Prandtl and Schmidt numbers differ from the value 1·0, it is still possible to obtain numerical solutions to equations (17.2.7) and (17.2.8) by inserting the appropriate value for the ratio $\nu/\mathscr{D}$ or ν/α and by taking the same form as before for the function f from the Blasius solution for the velocity distribution. The rate of transfer at the wall may then be calculated from the concentration or temperature profile, and hence the transfer coefficients may be derived for any specified value of Pr or Sc. These numerical results may be expressed *approximately* by

$$\frac{G_A}{u_1 \Delta c_A} = 0 \cdot 332 Sc^{-\frac{2}{3}} Re^{-\frac{1}{2}}, \qquad (17.2.10)$$

and

$$St = \frac{q}{\rho u_1 c_p \Delta T} = 0 \cdot 332 Pr^{-\frac{2}{3}} Re^{-\frac{1}{2}}. \qquad (17.2.11)$$

These results are for the local transfer coefficients at distance x from the leading edge. For the *average transfer coefficients* over a length x, the results are

$$\frac{G_A}{u_1 \Delta c_A} = 0 \cdot 664 Sc^{-\frac{2}{3}} Re^{-\frac{1}{2}}, \qquad (17.2.12)$$

and

$$St = 0 \cdot 664 Pr^{-\frac{2}{3}} Re^{-\frac{1}{2}}. \qquad (17.2.13)$$

In terms of a Nusselt number the last result becomes

$$Nu = 0 \cdot 664 Pr^{\frac{1}{3}} Re^{\frac{1}{2}}. \qquad (17.2.14)$$

Note that we have assumed the ordinary isothermal Blasius velocity distribution still applies when heat transfer is taking place. This is only true if temperature differences are small so that variations in the physical properties of the fluid may be neglected.

The result (17.2.13) for the heat-transfer coefficient should be compared with (16.4.8) which was obtained by the approximate methods of §16.4.

17.3. Forced convection with laminar flow in a pipe

Consider next the case of laminar flow in a circular pipe with mass or heat transfer taking place. Let the temperature of the fluid be T_1 at entry to the heated or cooled length of the pipe. It will be assumed that T_1 is constant across the flow. Let the pipe wall temperature be T_0 over the heated or cooled section. Let the temperature at any other point in the flow be T.

Writing $\theta = \dfrac{T - T_0}{T_1 - T_0}$ the energy equation (17.1.4a) becomes

$$\frac{D\theta}{Dt} = \alpha \nabla^2 \theta, \qquad (17.3.1)$$

and the boundary conditions which must be satisfied are

$$\theta = 0 \text{ at } r = a \text{ for all values of } z > 0,$$
$$\theta = 1 \text{ at } z = 0.$$

The problem is illustrated diagrammatically in fig. 90.

For the corresponding problem in mass transfer, c_{A_1} will be the concentration of the diffusing component (A) at the entry section $z = 0$, c_{A_0} will be the concentration at the wall, and c_A the value at any other point in the flow. Writing $\psi = \dfrac{c_A - c_{A_0}}{c_{A_1} - c_{A_0}}$ the mass-transfer equation (17.1.3 a) becomes

$$\frac{D\psi}{Dt} = \mathscr{D}\nabla^2\psi,$$

and the boundary conditions are the same as in the heat-transfer

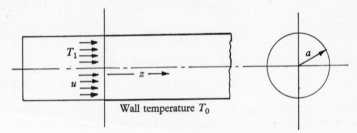

Wall temperature T_0

Fig. 90

problem. The solution of equation (17.3.1) will therefore apply equally to the mass-transfer problem if $\mathscr{D}$ is substituted for α.

In cylindrical polar coordinates $\nabla^2\theta = \dfrac{\partial^2\theta}{\partial r^2} + \dfrac{1}{r}\dfrac{\partial\theta}{\partial r} + \dfrac{\partial^2\theta}{\partial z^2}$. We will neglect longitudinal conduction, however, i.e. the term $\partial^2\theta/\partial z^2$ will be omitted. Equation (17.3.1) therefore becomes

$$u\frac{\partial\theta}{\partial z} = \alpha\left[\frac{\partial^2\theta}{\partial r^2} + \frac{1}{r}\frac{\partial\theta}{\partial r}\right]. \tag{17.3.2}$$

There are two special cases to be considered:

(a) Near the entrance to a pipe the velocity will be nearly uniform over the cross-section, i.e. $u = u_m = $ constant. This is referred to as 'rod-like flow' or 'plug flow', and from (17.3.2) the equation to be solved for the temperature distribution is

$$u_m\frac{\partial\theta}{\partial z} = \alpha\left[\frac{\partial^2\theta}{\partial r^2} + \frac{1}{r}\frac{\partial\theta}{\partial r}\right]. \tag{17.3.3}$$

(b) For fully developed laminar flow at some distance from the pipe entry we may assume that the parabolic Poiseuille velocity distribution applies, i.e. $u = 2u_m(1 - r^2/a^2)$, and the equation to be solved for the temperature distribution is

$$2u_m \left(1 - \frac{r^2}{a^2}\right) \frac{\partial \theta}{\partial z} = \alpha \left[\frac{\partial^2 \theta}{\partial r^2} + \frac{1}{r} \frac{\partial \theta}{\partial r}\right]. \qquad (17.3.4)$$

The solution of these equations is discussed in Appendix 6. In the case of *rod-like flow* with uniform velocity across the pipe the temperature distribution is given by

$$\theta = 2 \sum_{n=1}^{\infty} \frac{1}{\beta_n a} \frac{J_0(\beta_n r)}{J_1(\beta_n a)} \exp\left(-\frac{\alpha}{u_m} \beta_n^2 z\right), \qquad (17.3.5)$$

where J_0 and J_1 are Bessel functions of the first kind of order 0 and 1 respectively, and β_n is the nth root of the equation

$$J_0(\beta a) = 0.$$

For the average or bulk temperature T_2 of the stream at distance $z = L$, i.e. the temperature averaged over a cross-section with respect to the radius r, the result is

$$\theta_2 = \frac{T_2 - T_0}{T_1 - T_0} = 4 \sum_{n=1}^{\infty} \frac{1}{(\beta_n a)^2} \exp\left(-\frac{\alpha \beta_n^2}{u_m} L\right) \qquad (17.3.6)$$

or $$\frac{T_2 - T_1}{T_0 - T_1} = 1 - 4 \sum_{n=1}^{\infty} \frac{1}{(\beta_n a)^2} \exp\left(-\frac{\alpha \beta_n^2}{u_m} L\right),$$

and noting that $\beta_n a$ is dimensionless and that $u_m d/\alpha$ is the Péclet number Pe specified in terms of the pipe diameter d, the last result may be written with λ_n in place of $\beta_n a$,

$$\frac{T_2 - T_1}{T_0 - T_1} = 1 - 4 \sum_{n=1}^{\infty} \frac{1}{\lambda_n^2} \exp\left(-4\lambda_n^2 \frac{L}{d} \frac{1}{Pe}\right), \qquad (17.3.7)$$

and for the corresponding problem in mass transfer

$$\frac{c_{A_2} - c_{A_1}}{c_{A_0} - c_{A_1}} = 1 - 4 \sum_{n=1}^{\infty} \frac{1}{\lambda_n^2} \exp\left(-4\lambda_n^2 \frac{L}{d} \frac{\mathscr{D}}{u_m d}\right). \qquad (17.3.8)$$

The total rate of heat transfer over a length L is given by

$$Q = \rho u_m \frac{\pi d^2}{4} c_p (T_2 - T_1),$$

the average heat flux at the surface of the pipe will therefore be

$$q = \rho u_m \frac{d}{4L} c_p (T_2 - T_1),$$

and the Stanton number is given by

$$St = \frac{q}{\rho u_m c_p \Delta T} = \frac{d}{4L} \frac{(T_2 - T_1)}{\Delta T}.$$

If ΔT is taken as the arithmetic mean-temperature difference, i.e.

$$\Delta T = T_0 - \tfrac{1}{2}(T_1 + T_2) = (T_0 - T_1) - \tfrac{1}{2}(T_2 - T_1),$$

and hence, from (17.3.7)

$$\frac{T_2 - T_1}{\Delta T} = 2 \frac{(1 - 4S)}{(1 + 4S)}, \quad \text{where} \quad S = \sum_{n=1}^{\infty} \frac{1}{\lambda_n^2} \exp\left(-4\lambda_n^2 \frac{L}{d} \frac{1}{Pe}\right),$$

and for the Stanton number

$$St = \frac{1}{2L/d}\left(\frac{1 - 4S}{1 + 4S}\right), \tag{17.3.9}$$

i.e. the dimensionless heat-transfer coefficient is expressed as a function of the Péclet number and the ratio L/d.

The corresponding Nusselt number hd/k is given by

$$Nu = \frac{Pe}{2L/d}\left(\frac{1 - 4S}{1 + 4S}\right). \tag{17.3.10}$$

The numerical values of the coefficients λ_n in the series of terms

$$S = \sum_{n=1}^{\infty} \frac{1}{\lambda_n^2} \exp\left(-4\lambda_n^2 \frac{L}{d} \frac{1}{Pe}\right),$$

are given by the roots of the equation $J_0(\lambda) = 0$. The first six roots are

$$\lambda_1 = 2 \cdot 405, \quad \lambda_2 = 5 \cdot 520, \quad \lambda_3 = 8 \cdot 654, \quad \lambda_4 = 11 \cdot 792,$$
$$\lambda_5 = 14 \cdot 931, \quad \lambda_6 = 18 \cdot 071.$$

17.4. Forced convection in turbulent flow

In the absence of a comprehensive theory of turbulence which would permit the direct calculation of the Reynolds stresses and the corresponding heat- and mass-transfer quantities, recourse must be had to one or other of the analogies which are based on mixing length theory. The Reynolds and Taylor-Prandtl analogies have already been discussed in Chapters 11 and 15. Further refinements are possible, such as the following extension of the Taylor-Prandtl analogy which is due to Kármán.

In Chapter 14, §14.4, it was concluded that turbulent flow parallel to a smooth surface could be divided into three zones as follows:

$y^* < 5$ laminar sub-layer with transfer by molecular processes,

$5 < y^* < 30$ buffer zone with molecular and eddy transfer,

$y^* > 30$ turbulent zone with eddy transfer predominant,

the quantity y^* being defined by $y^* = \dfrac{y \sqrt{(\tau_0/\rho)}}{\nu}$.

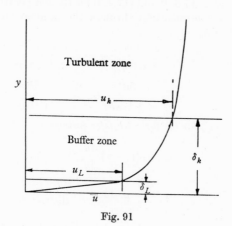

Fig. 91

For the velocity distribution, writing $u^* = \dfrac{u}{\sqrt{(\tau_0/\rho)}}$, we can take

in the laminar sub-layer $u^* = y^*$

in the buffer zone, using Kármán's approximate expression,
$$u^* = -3\cdot05 + 5\cdot0 \log_e y^*,$$

in the turbulent zone, from (14.4.3a), $u^* = 5\cdot5 + 2\cdot5 \log_e y^*$.

Note that the constants in Kármán's approximate expression for the velocity distribution in the buffer zone are chosen to give continuity both in velocity and velocity gradient at the edge of the laminar sub-layer where $y^* = 5$. At the junction between the buffer zone and the turbulent zone, however, these values give continuity only in the velocity and not in the velocity gradient. This outer edge of the buffer zone, specified by $y^* = 30$, however, has no real physical significance.

The situation regarding the transfer of momentum, heat and mass can be summarized as follows:

In the *laminar sub-layer*

$$\tau_0 = \mu \frac{du}{dy}, \tag{17.4.1}$$

$$q = -k \frac{dT}{dy}, \tag{17.4.2}$$

$$G_A = -\mathscr{D} \frac{dc_A}{dy}, \tag{17.4.3}$$

hence, integrating (17.4.2) and (17.4.3) for the temperature difference and change in concentration through the laminar sub-layer from $y^* = 0$ to $y^* = 5$,

$$T_L - T_0 = -\frac{5\nu}{k\sqrt{(\tau_0/\rho)}} q = -\frac{5Pr}{\sqrt{(\tau_0/\rho)}} \frac{q}{\rho c_p}, \tag{17.4.4}$$

and $\quad c_{A_L} - c_{A_0} = -\dfrac{5\nu}{\mathscr{D}\sqrt{(\tau_0/\rho)}} G_A = -\dfrac{5Sc}{\sqrt{(\tau_0/\rho)}} G_A. \tag{17.4.5}$

In the *buffer zone, for the three transfer processes*

$$\tau_0 = \rho(\nu + \varepsilon) \frac{d\bar{u}}{dy}, \tag{17.4.6}$$

$$q = -\rho c_p(\alpha + \varepsilon) \frac{dT}{dy}, \tag{17.4.7}$$

$$G_A = -(\mathscr{D} + \varepsilon) \frac{dc_A}{dy}, \tag{17.4.8}$$

where $\varepsilon = l'\tilde{v}'$ is the kinematic eddy viscosity. From the velocity profile for the buffer zone,

$$\frac{du^*}{dy^*} = \frac{5}{y^*} \quad \text{or} \quad \frac{du}{dy} = \frac{\tau_0}{\rho\nu} \frac{5}{y^*},$$

hence, from (17.4.6),

$$\left(1 + \frac{\varepsilon}{\nu}\right) = \frac{y^*}{5}$$

or $\qquad\qquad \varepsilon = \nu\left(\frac{y^*}{5} - 1\right), \tag{17.4.9}$

i.e. the kinematic eddy viscosity ranges from the value 0 at $y^* = 5$ to 5ν at the outer edge of the buffer zone where $y^* = 30$. Substituting for ε in (17.4.7) gives

$$\frac{q}{\rho c_p} = -\left(\alpha - \nu + \frac{\nu y^*}{5}\right) \frac{dT}{dy}$$

or
$$\frac{dT}{dy^*} = -\frac{q}{\rho c_p}\frac{Pr}{\sqrt{(\tau_0/\rho)}}\frac{1}{\left\{Pr\frac{y^*}{5} - (Pr - 1)\right\}}.$$

Hence, integrating from $y^* = 5$ to $y^* = 30$,

$$T_k - T_L = -\frac{q}{\rho c_p}\frac{5}{\sqrt{(\tau_0/\rho)}}\log_e(5Pr + 1), \qquad (17.4.10)$$

and similarly for mass transfer,

$$c_{A_k} - c_{A_L} = -G_A\frac{5}{\sqrt{(\tau_0/\rho)}}\log_e(5Sc + 1). \qquad (17.4.11)$$

For the velocity at the outer edge of the buffer zone we can say

$$u_k^* - u_L^* = 5{\cdot}0\log_e\tfrac{3{\cdot}0}{5} = 5{\cdot}0\log_e 6,$$

i.e.
$$u_k^* = 5(1 + \log_e 6) \qquad (17.4.12)$$

or
$$u_k = 5\sqrt{\frac{\tau_0}{\rho}}(1 + \log_e 6).$$

In the *turbulent zone*, extending beyond $y^* = 30$, we can assume that the Reynolds analogy applies for the movement of particles of fluid carrying mean properties u_m, T_m, c_{A_m}, up to the edge of the buffer zone. Hence, from (11.2.2) and (15.6.5),

$$\frac{q}{\tau_0} = -\frac{c_p(T_m - T_k)}{u_m - u_k}, \qquad (17.4.13)$$

and
$$\frac{G_A}{\tau_0} = -\frac{(c_{A_m} - c_{A_k})}{\rho(u_m - u_k)}. \qquad (17.4.14)$$

From (17.4.13)

$$T_m - T_k = -\frac{q}{\rho c_p}\frac{(u_m - u_k)}{\tau_0/\rho} = -\frac{q}{\rho c_p}\frac{(u_m^* - u_k^*)}{\sqrt{(\tau_0/\rho)}}.$$

Hence, substituting for u_k^* from (17.4.12), and noting that

$$\sqrt{\frac{\tau_0}{\rho}} = u_m\sqrt{\frac{c_f}{2}},$$

we have

$$T_m - T_k = -\frac{q}{\rho c_p}\frac{1}{\sqrt{(\tau_0/\rho)}}\left[\sqrt{\frac{2}{c_f}} - 5(1 + \log_e 6)\right], \quad (17.4.15)$$

and similarly for mass transfer

$$c_{A_m} - c_{A_k} = -\frac{G_A}{\sqrt{(\tau_0/\rho)}}\left[\sqrt{\frac{2}{c_f}} - 5(1 + \log_e 6)\right]. \quad (17.4.16)$$

Adding the three temperature differences from (17.4.4), (17.4.10) and (17.4.15) gives the result for the overall temperature difference

$$\Delta T = T_m - T_0 = -\frac{q}{\rho c_p}\frac{1}{\sqrt{(\tau_0/\rho)}}\left\{5Pr + 5\log_e(5Pr + 1)\right.$$

$$\left. + \sqrt{\frac{2}{c_f}} - 5(1 + \log_e 6)\right\},$$

and hence,

$$\Delta T = -\frac{q}{\rho c_p}\frac{1}{\sqrt{(\tau_0/\rho)}}\left\{5(Pr - 1)\right.$$

$$\left. + 5\log_e[1 + \tfrac{5}{6}(Pr - 1)] + \sqrt{\frac{2}{c_f}}\right\}$$

or, for the heat-transfer coefficient,

$$St = \frac{q}{\rho u_m c_p \Delta T} = \frac{\tfrac{1}{2}c_f}{1 + 5\sqrt{\dfrac{c_f}{2}}\left\{(Pr - 1) + \log_e[1 + \tfrac{5}{6}(Pr - 1)]\right\}}.$$

$$(17.4.17)$$

The corresponding result for mass transfer is

$$\frac{G_A}{u_m\Delta c_A} = \frac{\tfrac{1}{2}c_f}{1 + 5\sqrt{\dfrac{c_f}{2}}\left\{(Sc - 1) + \log_e[1 + \tfrac{5}{6}(Sc - 1)]\right\}}. \quad (17.4.18)$$

CHAPTER 18

COMPRESSIBLE FLOW IN PIPES AND NOZZLES

18.1. Compressible flow through a convergent nozzle

The one-dimensional frictionless flow of a perfect gas has been discussed previously in §4.7. The basic equations may be stated as follows:

Continuity   mass flow $m = \rho A u.$ (18.1.1)

Energy equation $H_0 = H + \tfrac{1}{2}u^2$ (18.1.2)

or $c_p T_0 = c_p T + \tfrac{1}{2}u^2,$ (18.1.2a)

and for the special case of *frictionless isentropic flow,*

$$\frac{p}{p_0} = \left(\frac{\rho}{\rho_0}\right)^{\gamma} = \left(\frac{T}{T_0}\right)^{\gamma/(\gamma-1)}.\qquad (18.1.3)$$

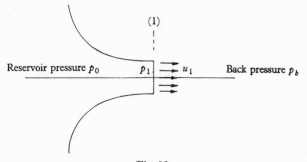

Fig. 92

Hence, from (18.1.2a) and (18.1.3), for frictionless isentropic flow from a reservoir, the velocity u at any point specified by the pressure p and temperature T is given by

$$\frac{u^2}{2} = c_p T_0 \left(1 - \frac{T}{T_0}\right) = \frac{\gamma}{\gamma-1}\frac{p_0}{\rho_0}\left[1 - \left(\frac{p}{p_0}\right)^{(\gamma-1)/\gamma}\right]. \quad (18.1.4)$$

Substituting for ρ and u in the continuity equation we have

$$m = \rho_0 \left(\frac{p}{p_0}\right)^{1/\gamma} \sqrt{\left(\frac{2\gamma}{\gamma-1}\frac{p_0}{\rho_0}\right)\left[1 - \left(\frac{p}{p_0}\right)^{(\gamma-1)/\gamma}\right]^{\frac{1}{2}}} A$$

or $\qquad m = \rho_0 a_0 A \sqrt{\dfrac{2}{\gamma - 1} \left[\left(\dfrac{p}{p_0}\right)^{2/\gamma} - \left(\dfrac{p}{p_0}\right)^{(\gamma + 1)/\gamma} \right]^{\frac{1}{2}}}, \qquad (18.1.5)$

where $\quad a_0 = \sqrt{\dfrac{\gamma p_0}{\rho_0}}.$

If A is the flow area at the exit section of the nozzle and p is the back pressure, equation (18.1.5) will give the mass flow m for any value

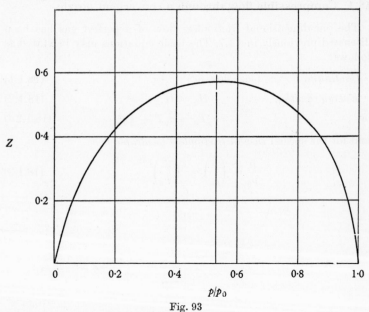

Fig. 93

of the outlet pressure p and for any shape of nozzle provided the flow is frictionless and isentropic.

Equation (18.1.5) may be written

$$m = \rho_0 a_0 A Z, \qquad (18.1.5a)$$

where the quantity Z is given by

$$Z = \sqrt{\dfrac{2}{\gamma - 1} \left[\left(\dfrac{p}{p_0}\right)^{2/\gamma} - \left(\dfrac{p}{p_0}\right)^{(\gamma + 1)/\gamma} \right]^{\frac{1}{2}}}.$$

This function is plotted against the pressure ratio p/p_0 in fig. 93 for the case of $\gamma = 1.4$. Similar curves can be plotted for other values of γ.

It will be of interest to consider the case of a *convergent* nozzle and the variation of the mass flow m with a progressive reduction in the back pressure p_b. Starting with a value of the back pressure p_b equal to the reservoir pressure p_0, the mass flow will be zero. If the back pressure is now reduced, the mass flow will increase steadily and will be given by equation (18.1.5). The quantity Z reaches a maximum, however, at the critical pressure ratio given by

$$\frac{p}{p_0} = \left(\frac{2}{\gamma + 1}\right)^{\gamma/(\gamma - 1)}, \tag{18.1.6}$$

and the mass flow m will therefore also reach its maximum value at this point. Under these conditions sonic velocity is attained at the exit section. Equation (18.1.5) and the curve of fig. 93 would imply that with further reduction of the back pressure below the critical value the mass flow would decrease. This does not in fact occur, however, because the condition of isentropic flow on which equation (18.1.5) is based is no longer satisfied beyond the throat. If the back pressure is reduced below the critical value, the nozzle is said to be *choked* and the pressure p_1 inside the nozzle at the exit section will in fact be the critical value given by equation (18.1.6). The flow then emerges from the nozzle as a free jet into a region of lower pressure p_b.

For a *convergent nozzle* therefore, if the back pressure is less than the critical value the discharge velocity will be given by

$$u_1 = \sqrt{\left(\frac{2\gamma}{\gamma + 1} \frac{p_0}{\rho_0}\right)}, \tag{18.1.7}$$

which is the critical velocity corresponding to $\dfrac{p_1}{p_0} = \left(\dfrac{2}{\gamma + 1}\right)^{\gamma/(\gamma - 1)}$. Note also that with isentropic conditions of flow up to (but not beyond) the exit section, $p_0/\rho_0^{\gamma} = p_1/\rho_1^{\gamma}$, and hence substituting the critical value for p_1 from (18.1.6),

$$\frac{p_0}{\rho_0} = \frac{p_1}{\rho_1}\left(\frac{p_0}{p_1}\right)^{(\gamma - 1)/\gamma} = \frac{(\gamma + 1)}{2}\frac{p_1}{\rho_1},$$

i.e. $u_1 = \sqrt{(\gamma p_1/\rho_1)}$, which is the sonic velocity at section (1). The mass flow is then given by

$$m = \rho_1 A_1 u_1 = \rho_1 A_1 \sqrt{\frac{\gamma p_1}{\rho_1}} = \rho_0 a_0 A_1 \left(\frac{p_1}{p_0}\right)^{(\gamma + 1)/2\gamma}$$

or $$m = \rho_0 a_0 A_1 \left(\frac{2}{\gamma + 1}\right)^{(\gamma + 1)/\{2(\gamma - 1)\}}, \tag{18.1.8}$$

but note carefully that this only applies when the nozzle is choked, and that with higher values of the back pressure the mass flow is given by (18.1.5).

18.2. Compressible flow through a convergent-divergent nozzle

The nature of the flow through a de Laval convergent-divergent nozzle is shown diagrammatically in fig. 94. Equations (18.1.1)–(18.1.5) will still apply provided the flow is frictionless and isentropic.

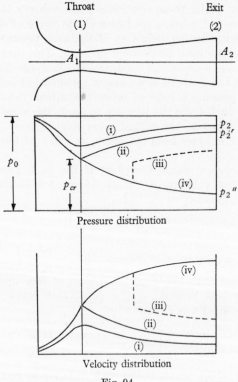

Fig. 94

Consider first of all the case where the critical pressure and sonic velocity are reached at the throat. The mass flow under isentropic conditions is given by

$$m = \rho_0 a_0 A Z,$$

where
$$Z = \sqrt{\frac{2}{\gamma - 1} \left[\left(\frac{p}{p_0}\right)^{2/\gamma} - \left(\frac{p}{p_0}\right)^{(\gamma + 1)/\gamma} \right]^{\frac{1}{2}}}.$$

To satisfy the equation of continuity, the values of the function Z at sections (1) and (2) must be related by

$$\frac{Z_1}{Z_2} = \frac{A_2}{A_1}.$$

It therefore follows from fig. 95, which shows the function Z plotted against the pressure ratio p/p_0, that with the maximum or critical value of Z applying at section (1) at the throat there are two possible values of p_2/p_0 corresponding to the value of Z_2 appropriate to the exit section (2). There are therefore two possible values for the pressure p_2 at the exit section for which isentropic shock-less flow can take place in the divergent portion of the nozzle while critical

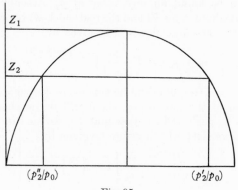

Fig. 95

conditions exist at the throat. These two values p'_2 and p''_2, as shown in fig. 94, correspond respectively to subsonic and shock-less supersonic flow in the divergent portion of the nozzle.

Four different cases of flow can now be distinguished corresponding to different values of the back pressure beyond the exit section. Starting with zero flow at $p_b = p_0$, let the back pressure be reduced progressively.

(i) If $p_b > p'_2$ we have subsonic flow throughout the nozzle,

the exit pressure	$p_2 = p_b$,

exit velocity	$u_2 = \sqrt{\left\{ \frac{2\gamma}{\gamma - 1} \frac{p_0}{\rho_0} \left[1 - \left(\frac{p_2}{p_0} \right)^{(\gamma - 1)/\gamma} \right] \right\}}$,

and the mass flow is given by equation (18.1.5), i.e.

$$m = \rho_0 a_0 A_2 \sqrt{\frac{2}{\gamma - 1} \left[\left(\frac{p_2}{p_0} \right)^{2/\gamma} - \left(\frac{p_2}{p_0} \right)^{(\gamma + 1)/\gamma} \right]^{\frac{1}{2}}}.$$

(ii) If $p_b = p_2'$ the flow will be just sonic at the throat but will immediately revert to subsonic conditions in the divergent portion. The above results for case (i) still apply, but we can say alternatively for the mass flow

$$m = \rho_0 a_0 A_1 \left(\frac{2}{\gamma+1}\right)^{(\gamma+1)/\{2(\gamma-1)\}}.$$

(iii) If $p_2' > p_b > p_2''$ the flow will be supersonic at first beyond the throat, but a shock wave will exist somewhere in the divergent portion of the nozzle and subsonic flow will therefore be re-established downstream from the shock wave. For any given location of the shock wave, the exit velocity u_2 and exit pressure p_2 may be calculated. Hence by a process of trial and error the theoretical exit velocity can be found for any value of p_2. Actually this case is somewhat idealized in fig. 94 and the real shock-wave pattern may be more complex. The mass flow will be given by

$$m = \rho_0 a_0 A_1 \left(\frac{2}{\gamma+1}\right)^{(\gamma+1)/\{2(\gamma-1)\}}.$$

(iv) If $p_b = p_2''$ the flow will be supersonic throughout the divergent portion of the nozzle and there will be no shock waves. If friction can be neglected the flow may be assumed to be isentropic and the exit velocity will therefore be given by

$$u_2 = \sqrt{\left/\left\{\frac{2\gamma}{\gamma-1}\frac{p_0}{\rho_0}\left[1 - \left(\frac{p_2}{p_0}\right)^{(\gamma-1)/\gamma}\right]\right\}\right.},$$

and the mass flow can be expressed either by

$$m = \rho_0 a_0 A_2 \sqrt{\frac{2}{\gamma-1}\left[\left(\frac{p_2}{p_0}\right)^{2/\gamma} - \left(\frac{p_2}{p_0}\right)^{(\gamma+1)/\gamma}\right]^{\frac{1}{2}}}$$

or, by $$m = \rho_0 a_0 A_1 \left(\frac{2}{\gamma+1}\right)^{(\gamma+1)/\{2(\gamma-1)\}}.$$

Further reduction of the back pressure below the value p_2'' will have no effect on the flow.

It follows from the last two results for case (iv) that if a de Laval nozzle is to be designed for supersonic flow in the divergent portion, the throat and exit areas A_1 and A_2 should be chosen in relation to the overall pressure ratio p_2/p_0, so that

$$m = \rho_0 a_0 A_1 \left(\frac{2}{\gamma+1}\right)^{(\gamma+1)/\{2(\gamma-1)\}}$$

$$= \rho_0 a_0 A_2 \sqrt{\frac{2}{\gamma-1}\left[\left(\frac{p_2}{p_0}\right)^{2/\gamma} - \left(\frac{p_2}{p_0}\right)^{(\gamma+1)/\gamma}\right]^{\frac{1}{2}}},$$

i.e. $\dfrac{A_1}{A_2} = \left(\dfrac{\gamma+1}{2}\right)^{(\gamma+1)/\{2(\gamma-1)\}} \sqrt{\dfrac{2}{\gamma-1}\left[\left(\dfrac{p_2}{p_0}\right)^{2/\gamma} - \left(\dfrac{p_2}{p_0}\right)^{(\gamma+1)/\gamma}\right]}\,^{\frac{1}{2}}$

or $\dfrac{A_1}{A_2} = \left(\dfrac{\gamma+1}{2}\right)^{1/(\gamma-1)} \left[\dfrac{\gamma+1}{\gamma-1}\left\{\left(\dfrac{p_2}{p_0}\right)^{2/\gamma} - \left(\dfrac{p_2}{p_0}\right)^{(\gamma+1)/\gamma}\right\}\right]^{\frac{1}{2}}.$

18.3. Shock waves

The case of the normal shock wave has been outlined previously in § 4.8. The three conditions of conservation of mass, momentum and energy must be satisfied. Using the symbol G for the mass flow per unit area, and referring to fig. 96, we can say

continuity $\qquad\qquad \rho u = G = \rho_1 u_1,$ $\qquad\qquad$ (18.3.1)

momentum $\qquad\qquad p + \rho u^2 = p_1 + \rho_1 u_1^2,$ $\qquad\qquad$ (18.3.2)

energy $\qquad\qquad H + \tfrac{1}{2}u^2 = H_1 + \tfrac{1}{2}u_1^2,$ $\qquad\qquad$ (18.3.3)

or, for a gas with $H = c_p T$

$$c_p T + \tfrac{1}{2}u^2 = c_p T_1 + \tfrac{1}{2}u_1^2. \qquad (18.3.3a)$$

From (18.3.1) and (18.3.2) the continuity and momentum equations require that

$$p + \frac{G^2}{\rho} = p_1 + \frac{G^2}{\rho_1}, \qquad (18.3.4)$$

which is the equation of a *Rayleigh line*.

From (18.3.1) and (18.3.3) the continuity and energy equations require that

$$H + \frac{G^2}{2\rho^2} = H_1 + \frac{G^2}{2\rho_1^2}, \qquad (18.3.5)$$

which is the equation of a *Fanno line*.

For a perfect gas (18.3.5) may be written

$$c_p T + \frac{G^2}{2\rho^2} = c_p T_1 + \frac{G^2}{2\rho_1^2} \qquad (18.3.5a)$$

or $\qquad \dfrac{\gamma}{\gamma-1}\dfrac{p}{\rho} + \dfrac{G^2}{2\rho^2} = \dfrac{\gamma}{\gamma-1}\dfrac{p_1}{\rho_1} + \dfrac{G^2}{2\rho_1^2}.$ $\qquad$ (18.3.5b)

Hence, from (18.3.4) and (18.3.5b) we have the *Rankine-Hugoniot* relationship between the pressures and densities at sections (1) and (2) on either side of a normal shock wave

$$\tfrac{1}{2}(p_2 - p_1)\left(\frac{1}{\rho_1} + \frac{1}{\rho_2}\right) = \frac{\gamma}{\gamma-1}\left(\frac{p_2}{\rho_2} - \frac{p_1}{\rho_1}\right), \qquad (18.3.6)$$

which can be expressed alternatively by

$$\frac{p_2}{p_1} = \frac{\dfrac{\gamma+1}{\gamma-1}\dfrac{\rho_2}{\rho_1} - 1}{\dfrac{\gamma+1}{\gamma-1} - \dfrac{\rho_2}{\rho_1}}, \qquad (18.3.6a)$$

It was shown in §4.8 that the ratio of the velocities on either side of the shock wave is given by

$$\frac{u_2}{u_1} = \frac{(\gamma-1)M_1^2 + 2}{(\gamma+1)M_1^2}, \qquad (18.3.7)$$

hence, from the continuity equation (18.3.1),

$$\frac{\rho_2}{\rho_1} = \frac{u_1}{u_2} = \frac{(\gamma+1)M_1^2}{(\gamma-1)M_1^2 + 2}, \qquad (18.3.8)$$

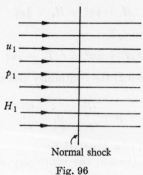

u_1

p_1

H_1

Normal shock

Fig. 96

and from the Rankine-Hugoniot relationship (18.3.6a)

$$\frac{p_2}{p_1} = \frac{2\gamma M_1^2 - (\gamma-1)}{(\gamma+1)}. \qquad (18.3.9)$$

The change of entropy of the fluid in passing through the shock wave is given by

$$s_2 - s_1 = c_p \log_e \frac{T_2}{T_1} - R \log_e \frac{p_2}{p_1}$$

or

$$s_2 - s_1 = c_v \log_e \frac{p_2}{p_1} + c_p \log_e \frac{\rho_1}{\rho_2},$$

hence, substituting from (18.3.8) and (18.3.9),

$$s_2 - s_1 = c_v \log_e \left\{ \frac{2\gamma M_1^2 - (\gamma-1)}{\gamma+1} \right\} + c_p \log_e \left\{ \frac{(\gamma-1)M_1^2 + 2}{(\gamma+1)M_1^2} \right\}.$$

The entropy difference $s_2 - s_1$ is positive if M_1 is greater than 1, i.e. a normal shock wave can only occur in a supersonic stream. In passing through the shock wave the flow must change from supersonic to subsonic velocity and from lower to higher pressure. The reverse process could not occur spontaneously because it would involve a decrease of entropy which would contradict the second law of thermodynamics.

In addition to normal shock waves, it is possible for *oblique shock waves* to occur. These are associated with a change in the direction of flow in a supersonic stream as indicated in fig. 97. The angle of the shock wave β is related to the angle of deflexion θ and to the

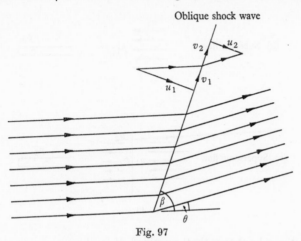

Fig. 97

incident Mach number M_1. It can be shown that, if the angle of deflexion θ is small the shock wave will be inclined at an angle approximating to the Mach angle $\sin^{-1} 1/M_1$.

18.4. Pitot tube in a high-velocity stream

With *subsonic flow* we may assume isentropic conditions, i.e.

$$\frac{p_0}{p_1} = \left(\frac{T_0}{T_1}\right)^{\gamma/(\gamma-1)},$$

and hence, from (4.7.5),

$$\frac{p_0}{p_1} = \left[1 + \frac{\gamma-1}{2} M_1^2\right]^{\gamma/(\gamma-1)}, \tag{18.4.1}$$

which may also be expressed in the form

$$\frac{p_0 - p_1}{\frac{1}{2}\rho u_1^2} = \frac{2}{\gamma M_1^2}\left\{\left[1 + \frac{\gamma-1}{2} M_1^2\right]^{\gamma/(\gamma-1)} - 1\right\}.$$

With *supersonic flow* a shock wave will occur ahead of the pitot tube as indicated in fig. 98 (*b*). Across the shock we will have the pressure ratio given by

$$\frac{p_2}{p_1} = \frac{2\gamma M_1^2 - (\gamma - 1)}{(\gamma + 1)}.$$

Beyond the shock the flow must be subsonic up to the mouth of the pitot tube, hence

$$\frac{p_3}{p_2} = \left[1 + \frac{\gamma - 1}{2} M_2^2\right]^{\gamma/(\gamma-1)}.$$

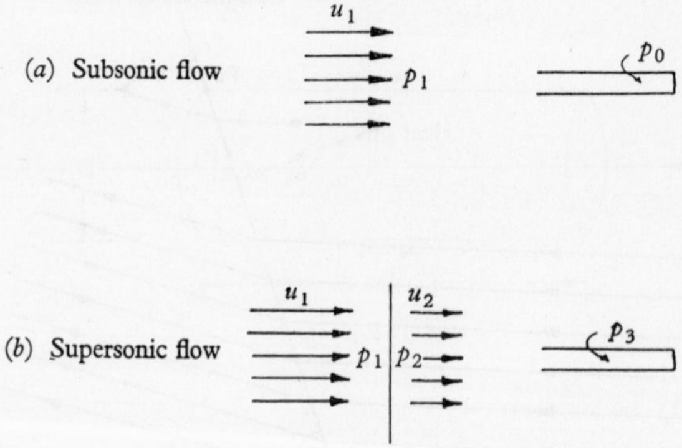

(*a*) Subsonic flow

(*b*) Supersonic flow

Fig. 98

Hence it can be shown that, provided $M_1 > 1$,

$$\frac{p_3}{p_1} = \left[\frac{(\gamma + 1)}{2} M_1^2\right]^{\gamma/(\gamma-1)} \Big/ \left[\frac{2\gamma M_1^2 - (\gamma - 1)}{(\gamma + 1)}\right]^{1/(\gamma-1)}. \quad (18.4.2)$$

18.5. Compressible flow in a pipe with friction

We can now consider the problem of the compressible flow of a gas through a pipe of uniform cross-section with skin friction and heat transfer. The following equations must apply:

continuity $\qquad\qquad \rho u = \text{constant} = G, \qquad\qquad (18.5.1)$

i.e. $\qquad\qquad \dfrac{du}{dx} = -\dfrac{G}{\rho^2}\dfrac{d\rho}{dx}; \qquad\qquad (18.5.1a)$

energy $\qquad\qquad G\,\dfrac{\pi D^2}{4}\,(dH + u\,du) = q\pi D\,dx,$

where q is the heat flux across the surface of the pipe, i.e.

$$dH + u\,du = \frac{4q}{GD}\,dx = \frac{4h\Delta T}{GD}\,dx,$$

where ΔT is the temperature difference between the wall of the pipe and the fluid. Hence

$$\frac{dH}{dx} + u\,\frac{du}{dx} = \frac{4St\,c_p\Delta T}{D} \qquad (18.5.2)$$

or, for a gas with $H = c_pT$,

$$c_p\,\frac{dT}{dx} + u\,\frac{du}{dx} = \frac{4St\,c_p\Delta T}{D}, \qquad (18.5.2a)$$

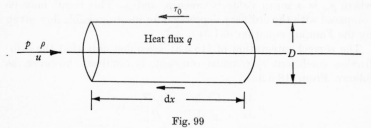

Fig. 99

which may be written

$$c_p\,\frac{dT}{dx} - \frac{G^2}{\rho^3}\,\frac{d\rho}{dx} = \frac{4St\,c_p\Delta T}{D}. \qquad (18.5.2b)$$

If the temperature difference ΔT is constant or a known function of the distance x measured along the axis of the pipe, equation $(18.5.2b)$ may be integrated to give a relationship between the temperature T of the fluid and the density ρ, i.e.

$$c_p(T - T_1) + \frac{G^2}{2}\left[\frac{1}{\rho^2} - \frac{1}{\rho_1^2}\right] = \frac{4St\,c_p}{D}\int_0^x \Delta T\,dx.$$

For the special case of adiabatic flow, i.e. with no heat transfer,

$$c_p(T - T_1) + \frac{G^2}{2}\left[\frac{1}{\rho^2} - \frac{1}{\rho_1^2}\right] = 0,$$

which is the equation of a Fanno line:

$$momentum \qquad -\frac{\pi D^2}{4}\,dp - \tau_0\pi D\,dx = \frac{\pi D^2}{4}\,G\,du,$$

i.e.
$$dp + G\,du + \frac{4\tau_0}{D}\,dx = 0$$

or
$$\frac{\mathrm{d}p}{\mathrm{d}x} + G\frac{\mathrm{d}u}{\mathrm{d}x} + \frac{2c_f G^2}{\rho D} = 0. \qquad (18.5.3)$$

Substituting from (18.5.1 a), the last equation may be written

$$\frac{\mathrm{d}p}{\mathrm{d}x} - \frac{G^2}{\rho^2}\frac{\mathrm{d}\rho}{\mathrm{d}x} + \frac{2c_f G^2}{\rho D} = 0. \qquad (18.5.3a)$$

If we take a mean value for the density ρ_m in the third term of this equation, integration will give the following approximate expression for the pressure drop in a length L of the pipe:

$$p_1 - p_2 = \frac{2c_f G^2}{\rho_m}\frac{L}{D} + G^2\left(\frac{1}{\rho_2} - \frac{1}{\rho_1}\right), \qquad (18.5.4)$$

where ρ_m is a mean value between ρ_1 and ρ_2. This result may be compared with the ordinary expression for incompressible flow given by the Fanning equation (6.1.3).

The correct integration of (18.5.3), assuming only that the skin-friction coefficient c_f remains constant, is obtained, however, as follows. From (18.5.3 a)

$$\rho\frac{\mathrm{d}p}{\mathrm{d}x} - \frac{G^2}{\rho}\frac{\mathrm{d}\rho}{\mathrm{d}x} + \frac{2c_f G^2}{D} = 0,$$

and hence,

$$\int_1^2 \rho\,\mathrm{d}p + G^2\log_e\frac{\rho_1}{\rho^2} + 2c_f G^2\frac{L}{D} = 0. \qquad (18.5.5)$$

The first term $\int_1^2 \rho\,\mathrm{d}p$ in equation (18.5.5) may be evaluated with the help of the equation of state, e.g. the perfect gas law

$$\frac{p}{\rho} = RT,$$

together with the energy equation (18.5.2), which for any particular case will contain the necessary information regarding the rate of heat supply to the fluid. In most cases, however, the solution is tedious.

The special case of *isothermal flow* in a pipe may be solved directly as follows. For a perfect gas at constant temperature we will have

$$\int_1^2 \rho\,\mathrm{d}p = \frac{\rho_1}{p_1}\int_1^2 p\,\mathrm{d}p = \frac{\rho_1}{p_1}\left(\frac{p_2^2 - p_1^2}{2}\right),$$

and hence, from (18.5.5),

$$\frac{p_1^2 - p_2^2}{2} = G^2\frac{p_1}{\rho_1}\left[2c_f\frac{L}{D} + \log_e\frac{\rho_1}{\rho_2}\right]. \qquad (18.5.6)$$

Noting that $\frac{1}{2}(p_1 + p_2)$ is the arithmetic mean pressure p_m, and that

$$\frac{p_1}{p_m \rho_1} = \frac{1}{\rho_m},$$

where ρ_m is the mean density, equation (18.5.6) may be written

$$p_1 - p_2 = \frac{G^2}{\rho_m}\left[2c_f \frac{L}{D} + \log_e \frac{\rho_1}{\rho_2}\right]. \qquad (18.5.6a)$$

For the special case of *adiabatic flow* in a pipe the energy equation takes the form of the equation for a Fanno line, i.e.

$$H + \frac{G^2}{2\rho^2} = \text{constant} = H_0, \qquad (18.5.7)$$

and for a perfect gas this may be written

$$\frac{\gamma}{\gamma - 1}\frac{p}{\rho} + \frac{G^2}{2\rho^2} = \frac{\gamma}{\gamma - 1}\frac{p_0}{\rho_0},$$

hence,

$$\frac{dp}{d\rho} = \frac{p_0}{\rho_0} + \frac{\gamma - 1}{2\gamma}\frac{G^2}{\rho^2},$$

and therefore

$$\int \rho\, dp = \int \left[\frac{p_0}{\rho_0}\rho + \frac{\gamma - 1}{2\gamma}\frac{G^2}{\rho}\right] d\rho$$

or

$$\int_1^2 \rho\, dp = \frac{p_0}{\rho_0}\frac{(\rho_2^2 - \rho_1^2)}{2} + \frac{\gamma - 1}{2\gamma}G^2 \log_e \frac{\rho_2}{\rho_1}.$$

Substituting for the first term in (18.5.5) we have the following relationship for the change of density with adiabatic flow in a pipe with friction:

$$\frac{p_0}{\rho_0}\frac{(\rho_2^2 - \rho_1^2)}{2} + \frac{(\gamma + 1)}{2\gamma}G^2 \log_e \frac{\rho_1}{\rho_2} + 2c_f G^2 \frac{L}{D} = 0. \qquad (18.5.8)$$

The pressure drop may then be evaluated from the Fanno equation, i.e.

$$\frac{\gamma}{\gamma - 1}\frac{p_1}{\rho_1} + \frac{G^2}{2\rho_1^2} = \frac{\gamma}{\gamma - 1}\frac{p_2}{\rho_2} + \frac{G^2}{2\rho_2^2}.$$

REFERENCES

(1) LIEPMANN and PUCKETT. *Aerodynamics of a Compressible Fluid* (Wiley).
(2) KEENAN. *Thermodynamics* (Wiley).
(3) SCHMIDT. *Thermodynamics* (Oxford).

CHAPTER 19

OPEN-CHANNEL FLOW

19.1. Non-uniform flow with friction

The case of uniform flow with friction in an open channel was discussed in §6.6. The method will now be extended to cover non-uniform flow, but we will restrict ourselves to cases of steady flow with only a *gradual variation* of depth and velocity profile with distance along the channel.

Provided the variation of depth λ with distance x is gradual the quantity $(p/\rho g + h)$ will be constant over a normal cross-section,

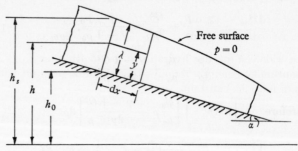

Fig. 100

although the total head will vary across the stream owing to variation of the velocity head. The total head at level y, where the pressure is p and the velocity u, is given by

$$H = \frac{u^2}{2g} + \frac{p}{\rho g} + h = \frac{u^2}{2g} + h_s,\dagger$$

or

$$H = \frac{u^2}{2g} + h_0 + \lambda \quad \text{if } \alpha \text{ is small.}$$

The mean total head for the stream is therefore approximately

$$H_m = \frac{\tilde{u}_m^2}{2g} + h_0 + \lambda, \tag{19.1.1}$$

where $\tilde{u}_m^2/2g$ is the mean value of $u^2/2g$.

† The symbol H is used to denote total head (not enthalpy) in this chapter.

The quantity $H_m - h_0$ is known as the specific head, i.e.

$$\text{specific head} = H_m - h_0 = \frac{\tilde{u}_m^2}{2g} + \lambda.$$

Alternatively we may define the specific energy E_s by

$$E_s = \frac{\tilde{u}_m^2}{2} + g\lambda.$$

From (19.1.1), assuming the ordinary mean velocity, u_m may be used in place of $\tilde{u}_m$:

$$\frac{\mathrm{d}H_m}{\mathrm{d}x} = \frac{u_m}{g}\frac{\mathrm{d}u_m}{\mathrm{d}x} + \frac{\mathrm{d}h_0}{\mathrm{d}x} + \frac{\mathrm{d}\lambda}{\mathrm{d}x},$$

and this must be equal to the loss of head per unit length due to friction, hence, comparing equation (6.6.2) or (6.6.3),

$$\frac{\mathrm{d}H_m}{\mathrm{d}x} = \frac{u_m}{g}\frac{\mathrm{d}u_m}{\mathrm{d}x} + \frac{\mathrm{d}h_0}{\mathrm{d}x} + \frac{\mathrm{d}\lambda}{\mathrm{d}x} = -\frac{c_f u_m^2}{2gm}, \qquad (19.1.2)$$

and for a wide channel the hydraulic mean depth $m = \lambda$.

The continuity equation requires that the flow per unit width of the stream must be constant, i.e.

$$\frac{Q}{b} = \lambda u_m = \text{constant} \qquad (19.1.3)$$

or $\qquad\qquad \dfrac{\mathrm{d}\lambda}{\lambda} + \dfrac{\mathrm{d}u_m}{u_m} = 0$ assuming b is constant,

hence, $\qquad \dfrac{u_m}{g}\dfrac{\mathrm{d}u_m}{\mathrm{d}x} = -\dfrac{u_m^2}{g\lambda}\dfrac{\mathrm{d}\lambda}{\mathrm{d}x} = -\dfrac{Q^2}{gb^2\lambda^3}\dfrac{\mathrm{d}\lambda}{\mathrm{d}x},$

and therefore from (19.1.2)

$$\left(1 - \frac{Q^2}{gb^2\lambda^3}\right)\frac{\mathrm{d}\lambda}{\mathrm{d}x} = -\frac{\mathrm{d}h_0}{\mathrm{d}x} - \frac{c_f u_m^2}{2g\lambda},$$

i.e. $\qquad \left(1 - \dfrac{Q^2}{gb^2\lambda^3}\right)\dfrac{\mathrm{d}\lambda}{\mathrm{d}x} = s_0 - \dfrac{c_f Q^2}{2gb^2\lambda^3}$

or $\qquad \dfrac{\mathrm{d}\lambda}{\mathrm{d}x} = \left\{s_0 - \dfrac{c_f}{2g}\dfrac{Q^2}{b^2\lambda^3}\right\}\bigg/\left\{1 - \dfrac{Q^2}{gb^2\lambda^3}\right\}, \qquad (19.1.4)$

where $s_0 = -\mathrm{d}h_0/\mathrm{d}x$ is the slope of the channel floor. The last equation may alternatively be written

$$\frac{\mathrm{d}\lambda}{\mathrm{d}x} = \left\{s_0 - \frac{c_f}{2g}\frac{u_m^2}{\lambda}\right\}\bigg/\left\{1 - \frac{u_m^2}{g\lambda}\right\}. \qquad (19.1.4a)$$

If $\lambda = c_f u_m^2/2gs_0$, $\mathrm{d}\lambda/\mathrm{d}x$ will be zero and we then have the case of uniform flow in the channel. The depth λ_0 for uniform flow given by

$$\lambda_0 = \frac{c_f u_m^2}{2gs_0} \quad \text{or} \quad \lambda_0 = \left(\frac{c_f Q^2}{2gs_0 b^2}\right)^{\frac{1}{3}}$$

is known as the *normal depth*.

Note also that $\quad \dfrac{\mathrm{d}\lambda}{\mathrm{d}x} \to \infty \quad$ as $\quad u_m \to \sqrt{(g\lambda)}$.

$\sqrt{(g\lambda)}$ is the velocity of propagation of a surface wave on the fluid in the channel. It may be regarded as a critical velocity for open-channel flow analogous to sonic velocity with compressible flow.

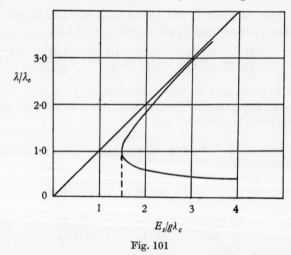

Fig. 101

For the case of a wide or rectangular channel, substituting from (19.1.3) for the mean velocity, the specific energy is given by

$$E_s = \frac{Q^2}{2b^2\lambda^2} + g\lambda \tag{19.1.5}$$

and $\quad \dfrac{\mathrm{d}E_s}{\mathrm{d}\lambda} = -\dfrac{Q^2}{b^2\lambda^3} + g \quad$ for constant flow Q,

hence the minimum specific energy occurs when $\lambda = (Q^2/b^2g)^{\frac{1}{3}}$. This is known as the critical depth λ_c, i.e.

$$\lambda_c = \left(\frac{Q^2}{b^2 g}\right)^{\frac{1}{3}}, \tag{19.1.6}$$

and the critical velocity is given by

$$u_{m_c} = \frac{Q}{b\lambda_c} = \left(\frac{gQ}{b}\right)^{\frac{1}{3}} = \sqrt{(g\lambda_c)}, \qquad (19.1.7)$$

which is identical with the critical velocity defined above. Hence

$$E_s = \frac{\lambda_c^3 g}{2\lambda^2} + g\lambda$$

or

$$\frac{E_s}{g\lambda_c} = \frac{1}{2}\frac{\lambda_c^2}{\lambda^2} + \frac{\lambda}{\lambda_c}. \qquad (19.1.8)$$

This relationship may be plotted as a specific energy diagram as in fig. 101. The minimum value of the specific energy at $\lambda = \lambda_c$ is

$$E_s = \tfrac{3}{2}g\lambda_c.$$

Flow with depth greater than the critical value λ_c and with velocities below the critical value $\sqrt{(g\lambda)}$ is described as *tranquil*. Flow with depth less than λ_c and with velocities greater than $\sqrt{(g\lambda)}$ is described as *rapid*.

19.2. Flow over a weir

Flow over a broad-crested weir is illustrated diagrammatically in fig. 102. Neglecting frictional effects, the specific energy of the

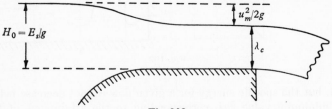

Fig. 102

stream is equal to gH_0, where H_0 is the head of the free surface at the reservoir measured above the crest of the weir. For a rectangular channel, from the definition of specific energy

$$E_s = \frac{Q^2}{2b^2\lambda^2} + g\lambda, \qquad (19.2.1)$$

i.e.

$$Q = \sqrt{2}\, b\lambda(E_s - g\lambda)^{\frac{1}{2}}. \qquad (19.2.2)$$

For maximum discharge at constant specific energy, $dQ/d\lambda = 0$, i.e.

$$\sqrt{2}\, b(E_s - g\lambda)^{\frac{1}{2}} - \frac{\sqrt{2}}{2} gb\lambda(E_s - g\lambda)^{-\frac{1}{2}} = 0,$$

and hence, the depth for maximum discharge $= \frac{2}{3}E_s/g$, which is the same value as the critical depth λ_c for minimum specific energy at constant Q.

If loss of specific energy due to friction can be neglected, therefore, the maximum discharge over the weir will be given by

$$Q_{\text{max.}} = \sqrt{2}\, b\lambda_c(E_s - g\lambda_c)^{\frac{1}{2}} = b\sqrt{g}\, \lambda_c^{\frac{3}{2}}. \qquad (19.2.3)$$

With negligible friction we can say $E_s = gH_0$, and hence

$$\lambda_c = \tfrac{2}{3}H_0.$$

The expression for the maximum discharge may therefore be written

$$Q_{\text{max.}} = (\tfrac{2}{3})^{\frac{3}{2}}b\sqrt{g}\, H_0^{\frac{3}{2}},$$

which is the result given by equation (2.7.10).

In practice, however, the effects of friction cannot be neglected entirely. Friction causes a steady fall in the specific energy of the

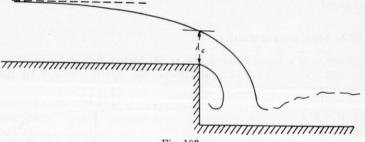

Fig. 103

flow, but the specific energy for a given flow cannot decrease below the minimum value $\frac{3}{2}g\lambda_c$ corresponding to the attainment of the critical depth λ_c. The flow over the weir will consequently tend to adjust itself so that the critical depth is just reached at the end as indicated in fig. 102. The discharge Q is therefore given correctly by equation (19.2.3), but the specific energy at the end of the weir will be less than gH_0 and the discharge will consequently be less than the theoretical value given by (2.7.10).

Consider next the problem of flow in a long channel ending in a fall as shown in fig. 103. From the definition of specific energy, and from equation (19.1.2) for the loss of head due to friction with gradually varying flow, we can say

$$\frac{1}{g}\frac{\mathrm{d}E_s}{\mathrm{d}x} + \frac{\mathrm{d}h_0}{\mathrm{d}x} = -\frac{c_f u_m^2}{2g\lambda}, \qquad (19.2.4)$$

or for a horizontal channel with $\dfrac{\mathrm{d}h_0}{\mathrm{d}x} = 0$

$$\frac{\mathrm{d}E_s}{\mathrm{d}x} = -\frac{c_f u_m^2}{2\lambda}.$$

Assuming that steady conditions of flow have been established, the last equation implies a steady decrease of the specific energy with distance along the channel. The flow rate will therefore adjust itself to a value Q so that the critical depth λ_c and the minimum value of the specific energy for this rate of flow are reached at the fall, i.e.

$$\lambda_c = \left(\frac{Q^2}{b^2 g}\right)^{\frac{1}{3}}.$$

19.3. Hydraulic jump

When a rapid flowing stream (i.e. with velocity above the critical value) has its velocity reduced to below the critical value, owing to

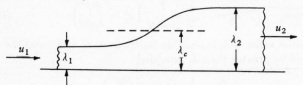

Fig. 104

an obstruction in the channel or to a change of slope, a surface discontinuity occurs which is known as the *hydraulic jump*. The phenomenon is in some ways analogous to a shock wave with compressible flow.

Consider the case of a horizontal channel of uniform width b. Referring to fig. 104, the following conditions must apply:

Continuity $Q = b\lambda_1 u_1 = b\lambda_2 u_2,$

i.e. $\lambda_1 u_1 = \lambda_2 u_2.$

Momentum, taking a control surface extending between sections (1) and (2):

force on section (1) due to fluid pressure $= \left(\rho g\,\dfrac{\lambda_1}{2}\right) b\lambda_1 = \rho g b\,\dfrac{\lambda_1^2}{2},$

force on section (2) $= \rho g b\,\dfrac{\lambda_2^2}{2},$

resultant force $= \dfrac{\rho g b}{2}\,(\lambda_1^2 - \lambda_2^2) =$ net rate of outflow of momentum

$$= \rho b\lambda_2 u_2^2 - \rho b\lambda_1 u_1^2,$$

i.e.
$$g \frac{(\lambda_1^2 - \lambda_2^2)}{2} = \lambda_2 u_2^2 - \lambda_1 u_1^2$$

$$= \frac{Q^2}{b^2} \left(\frac{1}{\lambda_2} - \frac{1}{\lambda_1} \right)$$

or
$$g \frac{(\lambda_1^2 - \lambda_2^2)}{2} = \frac{Q^2}{b^2} \frac{(\lambda_1 - \lambda_2)}{\lambda_1 \lambda_2},$$

and therefore to satisfy the equations of continuity and momentum we must have *either*

$$(\lambda_1 - \lambda_2) = 0, \quad \text{i.e. } \lambda_1 = \lambda_2, \text{ so that nothing happens,}$$

or
$$\lambda_1 \lambda_2 (\lambda_1 + \lambda_2) = \frac{2Q^2}{gb^2},$$

in which case

$$\lambda_2^2 + \lambda_1 \lambda_2 - \frac{2Q^2}{\lambda_1 gb^2} = 0,$$

and hence,
$$\lambda_2 = -\frac{\lambda_1}{2} \pm \frac{1}{2} \sqrt{\left(\lambda_1^2 + \frac{8Q^2}{\lambda_1 gb^2} \right)}, \qquad (19.3.1)$$

one positive value of λ_2 is possible (apart from $\lambda_2 = \lambda_1$) and the value of λ_2 will be greater than λ_1 provided

$$-\frac{\lambda_1}{2} + \frac{1}{2} \sqrt{\left(\lambda_1^2 + \frac{8Q^2}{\lambda_1 gb^2} \right)} > \lambda_1,$$

i.e.
$$\frac{1}{2} \sqrt{\left(\lambda_1^2 + \frac{8Q^2}{\lambda_1 gb^2} \right)} > \frac{3}{2} \lambda_1$$

or
$$\lambda_1^2 + \frac{8Q^2}{\lambda_1 gb^2} > 9\lambda_1^2,$$

i.e.
$$\left(\frac{Q^2}{gb^2} \right)^{\frac{1}{3}} > \lambda_1.$$

Noting that $\lambda_c = (Q^2/gb^2)^{\frac{1}{3}}$ the last condition implies that λ_1 must be less than λ_c for a hydraulic jump to occur with λ_2 greater than λ_1.

The loss of energy in the jump is given by

$$\left(\frac{u_1^2}{2} + g\lambda_1 \right) - \left(\frac{u_2^2}{2} + g\lambda_2 \right) = \tfrac{1}{2}(u_1^2 - u_2^2) + g(\lambda_1 - \lambda_2)$$

$$= \frac{u_1^2}{2} \left(1 - \frac{\lambda_1^2}{\lambda_2^2} \right) - g(\lambda_2 - \lambda_1)$$

$$= \frac{Q^2}{2b^2\lambda_1^2} \left(1 - \frac{\lambda_1^2}{\lambda_2^2} \right) - g(\lambda_2 - \lambda_1),$$

and noting that
$$\frac{Q^2}{b^2} = \frac{g}{2}\,\lambda_1\lambda_2(\lambda_1 + \lambda_2),$$

the loss of energy $= \dfrac{g}{4}\,\lambda_1\lambda_2(\lambda_1 + \lambda_2)\,\dfrac{(\lambda_2^2 - \lambda_1^2)}{\lambda_1^2\lambda_2^2} - g(\lambda_2 - \lambda_1),$

and hence, simplifying the last expression,

$$\text{loss of energy} = \frac{g}{4}\frac{(\lambda_2 - \lambda_1)^3}{\lambda_1\lambda_2}, \tag{19.3.2}$$

i.e. a loss of energy occurs provided $\lambda_2 > \lambda_1$, which confirms the previous statement that a jump can only occur from a value of λ less than the critical value λ_c to another value greater than λ_c. Compare this argument with that for compressible flow through a normal shock wave where flow can only occur in a direction from lower to higher pressure. The hydraulic jump is the usual mechanism by which a rapid flowing stream can revert to tranquil flow.

19.4. Flow below a sluice gate

A typical case of the occurrence of the hydraulic jump is found with flow below a sluice gate. If the stream emerges in rapid flow and if the conditions further downstream require the existence of tranquil flow, a jump will occur somewhere below the sluice gate. Two examples of this are illustrated in fig. 105. In case (a) the stream emerges with depth λ_1 at the vena contracta in a channel of mild slope where steady uniform flow is maintained at the *normal depth* λ_0 given by

$$\lambda_0 = \frac{c_f u_m^2}{2gs_0} = \left(\frac{c_f}{2gs_0}\frac{Q^2}{b^2}\right)^{\frac{1}{3}}.$$

The rapid-flowing stream will have a surface profile in this case as shown in fig. 105(a). For every point on this profile there must be a corresponding conjugate point to which the surface of the stream could jump. The locus of these conjugate points is shown by the dotted curve on the figure. The actual location of the jump will be determined by the point of intersection of the conjugate profile with the line for uniform flow at normal depth. If the normal depth λ_0 in case (a) is greater than the conjugate depth corresponding to the value λ_1 at the vena contracta, a free outflow could not be maintained since the jump would move back to the sluice gate and the flow would be submerged.

In case (b) a similar type of flow is illustrated, but, instead of uniform flow in the channel, we have the case of steady flow with friction in a long channel ending in a fall where the critical depth λ_c must be maintained. The form of the surface profile below the jump is typical of that for tranquil flow with depths intermediate between λ_0 and λ_c in a channel of mild slope. Note that the definition of *mild slope* is one for which the normal depth is greater than the

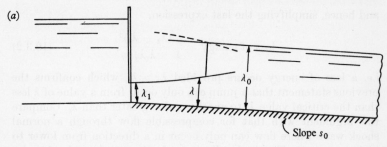

(a)

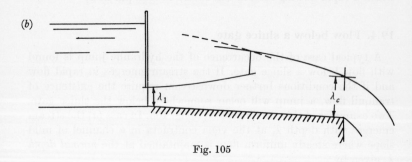

(b)

Fig. 105

critical depth. If, on the other hand, the normal depth is less than the critical depth, the slope is described as being *steep*.

REFERENCES

(1) BAKHMETEFF. *Hydraulics of Open Channels* (McGraw-Hill).
(2) ROUSE. *Fluid Mechanics for Hydraulic Engineers* (McGraw-Hill).

CHAPTER 20

SOLID PARTICLES IN FLUID FLOW

20.1. Shape factors and mean diameters

Consider a particle of irregular shape having a volume V and a surface area A. It is useful to take as a standard of comparison the *equivalent-volume sphere*, i.e. a sphere whose diameter D_s is given by

$$\frac{\pi}{6} D_s^3 = V$$

or

$$D_s = \left(\frac{6V}{\pi}\right)^{\frac{1}{3}}. \tag{20.1.1}$$

D_s is sometimes referred to as the *nominal diameter* of the particle. The surface area of the equivalent-volume sphere is $A_s = \pi D_s^2$. The *particle shape factor* λ is then defined by $\lambda = A/A_s$, i.e. λ is the ratio of the surface area of the particle to that of the equivalent-volume sphere. The quantity $1/\lambda$ is sometimes referred to as the *sphericity*. For the special case of a spherical particle $\lambda = 1$, but for all other shapes λ will be greater than 1.

If the particle diameter is specified in some other way (e.g. by the maximum linear dimension of the particle) and denoted by D_p, the volume V and surface area A must be expressed by

$$V = \phi_v \frac{\pi}{6} D_p^3, \quad \text{where } \phi_v \text{ is a volume factor}$$

and

$$A = \phi_a \pi D_p^2, \quad \text{where } \phi_a \text{ is a surface factor}.$$

Note that

$$\phi_v = \left(\frac{D_s}{D_p}\right)^3 \quad \text{and} \quad \phi_a = \lambda \left(\frac{D_s}{D_p}\right)^2.$$

In chemical engineering one usually encounters *mixtures of particles* of different sizes. Consider first of all the case of a mixture of N particles of similar shape but of different diameters specified by D_p. If we could plot the percentage (by number) of particles less than a stated size against stated size the result would be a size distribution curve such as that shown in fig. 106. The *median diameter* is defined as the particle diameter for which 50% of the particles in the mixture are less than the stated size.

Other mean diameters for a mixture of particles may be defined as follows:

Arithmetic mean diameter $\qquad D_{am} = \dfrac{\Sigma D_p}{N}.$

Geometric mean diameter $\qquad D_{gm} = \sqrt[N]{(D_{p_1} D_{p_2} \ldots D_{p_N})}$

or $\qquad \log_e D_{gm} = \dfrac{\Sigma(\log D_p)}{N}.$

Surface mean diameter $\qquad D_{sm}^2 = \dfrac{\Sigma D_p^2}{N}.$

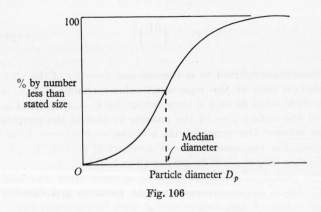

Fig. 106

Volume mean diameter $\qquad D_{vm}^3 = \dfrac{\Sigma D_p^3}{N}.$

Surface-diameter mean diameter $\qquad D_{sdm} = \dfrac{\Sigma D_p^2}{\Sigma D_p} = \dfrac{D_{sm}^2}{D_{am}}.$

Volume-diameter mean diameter $\qquad D_{vdm}^2 = \dfrac{\Sigma D_p^3}{\Sigma D_p} = \dfrac{D_{vm}^3}{D_{am}}.$

Volume-surface mean diameter $\qquad D_{vsm} = \dfrac{\Sigma D_p^3}{\Sigma D_p^2} = \dfrac{D_{vm}^3}{D_{sm}^2}.$

The surface mean diameter D_{sm} for instance is the diameter of a particle having a surface area equal to the average for all the particles in the mixture. The volume mean D_{vm} is the diameter of a particle having the average volume. The surface-diameter mean value D_{sdm} is the diameter of a particle having a ratio of surface to diameter equal to the average for the mixture. The volume-surface mean diameter D_{vsm} is the diameter of a particle having a ratio of volume to surface equal to the average for the mixture.

20.2. Particle-size distribution

If the particle-size distribution for a mixture followed the normal Gaussian or error distribution, we would define the deviation from the mean diameter by $x = D_p - D_{am}$, where D_{am} is the ordinary arithmetic mean value which for a Gaussian distribution is also the most probable diameter. We would then have for the number of particles with size deviation between x and $x + dx$,

$$dN = \frac{Nh}{\sqrt{\pi}}\, e^{-h^2x^2}\, dx, \tag{20.2.1}$$

where N is the total number of particles and h is a constant for a particular distribution. A high value for h implies a high proportion of particles around the mean size, i.e. the quantity $1/h$ gives a measure of the dispersion or scatter of sizes.

With a symmetrical Gaussian distribution the number of particles bigger than a stated size D_p, i.e. with values of the deviation x greater than $x = D_p - D_{am}$, is obtained by integrating (20.2.1), i.e. number of particles larger than size D_p

$$= \frac{Nh}{\sqrt{\pi}} \int_x^\infty e^{-h^2x^2}\, dx = \frac{N}{2}\left[1 - \mathrm{erf}\,(hx)\right].\dagger$$

The number of particles smaller than D_p would be given by

$$\frac{Nh}{\sqrt{\pi}} \int_{-\infty}^x e^{-h^2x^2}\, dx = \frac{N}{2}\left[1 + \mathrm{erf}\,(hx)\right],$$

and putting $x = 0$ the number of particles greater than D_{am} in size would be $\frac{1}{2}N$, which must obviously be the case with a symmetrical distribution. Also the median diameter by definition would be equal to the arithmetic mean diameter.

The *average deviation* is defined as the arithmetic mean of all the deviations taken without regard to sign. For a Gaussian distribution the average deviation a is given by $a = 1/h\sqrt{\pi}$.

The r.m.s. or *standard deviation* σ is defined by

$$\sigma^2 = \frac{\Sigma(x^2)}{N},$$

and for a Gaussian distribution the standard deviation $\sigma = 1/h\sqrt{2}$.

† The error function erf x is defined by

$$\mathrm{erf}\,x = \frac{2}{\sqrt{\pi}} \int_0^x e^{-x^2}\, dx.$$

A symmetrical or Gaussian distribution for the particle size in a mixture, however, would imply negative values for some particle diameters, which is obviously impossible. A more plausible distribution would be a *Gaussian distribution for* log D_p, and this is found to fit the facts with many practical cases of particle mixtures. We therefore define the size deviation by

$$x = \log_e D_p - \log_e D_{gm} \tag{20.2.2}$$

or

$$D_p = e^x D_{gm},$$

where D_{gm} is the geometric mean diameter. Note that when $D_p \to 0$, $x \to -\infty$.

A normal or Gaussian distribution for x as defined by (20.2.2) thus gives a logarithmic or 'skew probability' distribution for the particle size D_p.

The number of particles less than a stated size D_p is then given by

$$n = \frac{Nh}{\sqrt{\pi}} \int_{-\infty}^{x} e^{-h^2 x^2} \, dx \tag{20.2.3}$$

or

$$n = \tfrac{1}{2}N[1 + \text{erf} \, (hx)],$$

where x is defined by (20.2.2).

In practice it is not generally possible to measure the number of particles directly, but it is relatively easy to measure the mass of the particles in a given size range. Particle-size analysis is normally carried out by sieving and measuring the mass of particles having a diameter greater than or less than a stated size. For a logarithmic probability distribution of particle size the mass of particles in the size range x to $x + dx$ is given by

$$dM = \rho \phi_v \frac{\pi}{6} D_p^3 \frac{Nh}{\sqrt{\pi}} e^{-h^2 x^2} \, dx,$$

where the deviation x is defined by (20.2.2). Hence

$$dM = \rho \phi_v \frac{\pi}{6} D_{gm}^3 e^{3x} \frac{Nh}{\sqrt{\pi}} e^{-h^2 x^2} \, dx$$

$$= \rho \phi_v \frac{\pi}{6} D_{gm}^3 e^{9/4h^2} \frac{Nh}{\sqrt{\pi}} \exp\left\{ -h^2 \left(x - \frac{3}{2h^2} \right)^2 \right\} \, dx$$

or

$$dM = \rho \phi_v \frac{\pi}{6} D_{gm}^3 e^{9/4h^2} \frac{Nh}{\sqrt{\pi}} e^{-h^2 y^2} \, dy, \tag{20.2.4}$$

where

$$y = x - \frac{3}{2h^2}.$$

The total mass of all the particles will be given by

$$M = \rho\phi_v \frac{\pi}{6} D_{gm}^3 e^{9/4h^2} \frac{Nh}{\sqrt{\pi}} \int_{-\infty}^{+\infty} e^{-h^2 y^2} \, \mathrm{d}y$$

or $$M = \rho\phi_v \frac{\pi}{6} D_{gm}^3 e^{9/4h^2} N.$$

The *mass fraction* in the size range x to $x + \mathrm{d}x$ will therefore be given by

$$\frac{\mathrm{d}M}{M} = \frac{h}{\sqrt{\pi}} e^{-h^2 y^2} \, \mathrm{d}y. \tag{20.2.5}$$

The mass fraction of particles smaller than D_p in size will then be

$$\frac{h}{\sqrt{\pi}} \int_{-\infty}^{y} e^{-h^2 y^2} \, \mathrm{d}y = \tfrac{1}{2}[1 + \mathrm{erf}\,(hy)]. \tag{20.2.6}$$

Note that the median diameter for the mass distribution, i.e. the particle diameter for which the mass fraction of smaller particles is $\tfrac{1}{2}$, is determined by $\mathrm{erf}\,hy = 0$ and hence $y = 0$ and $x = \tfrac{3}{2}h^2$, i.e. the *median diameter for the mass distribution* D_{mm} is given by

$$D_{mm} = e^{3/2h^2} D_{gm}. \tag{20.2.7}$$

The *volume mean diameter* D_{vm} may be calculated in terms of D_{gm} as follows:

total mass $$M = N\rho\phi_v \frac{\pi}{6} D_{vm}^3 = \rho\phi_v \frac{\pi}{6} D_{gm}^3 e^{9/4h^2} N,$$

hence $$D_{vm}^3 = D_{gm}^3 e^{9/4h^2}$$

or $$D_{vm} = e^{3/4h^2} D_{gm} = e^{-3/4h^2} D_{mm}. \tag{20.2.8}$$

The following results may also be derived for the logarithmic probability distribution

Surface mean diameter $\qquad D_{sm} = e^{1/2h^2} D_{gm} = e^{-1/h^2} D_{mm}.$

Arithmetic mean diameter $\qquad D_{am} = e^{1/4h^2} D_{gm} = e^{-5/4h^2} D_{mm}.$

Volume-surface mean diameter $\quad D_{vsm} = e^{5/4h^2} D_{gm} = e^{-1/4h^2} D_{mm}.$

Another particle-size distribution is the *Rosin-Rammler relationship*, which is frequently used for the analysis of crushed solids and of liquid sprays. In this relationship a size deviation or ratio z is defined by

$$z = \frac{D_p}{D_{rm}},$$

where D_{rm} is a mean diameter which is yet to be specified. The *mass fraction* in the size range z to $z + dz$ is then expressed by

$$\frac{dM}{M} = kz^{k-1}e^{-z^k}\,dz, \qquad (20.2.9)$$

which takes the place of (20.2.5) for the logarithmic probability distribution. The constant k is a distribution constant. The mass fraction of particles smaller than D_p in size will then be given by

$$\int \frac{dM}{M} = \int_0^z kz^{k-1}e^{-z^k}\,dz = [1 - e^{-z^k}], \qquad (20.2.10)$$

which takes the place of (20.2.6) for the logarithmic probability distribution.

From (20.2.10) the mass fraction of particles smaller than D_{rm} in size will be given by

$$\int_{z=0}^{z=1} \frac{dM}{M} = 1 - e^{-1} = 0 \cdot 632,$$

and this leads to the definition of D_{rm} as the particle diameter for which the mass fraction of smaller particles is $0 \cdot 632$.

20.3. Solid particles in fluid streams

The behaviour of small *spherical* particles in a fluid stream can be predicted with reasonable accuracy. The motion of small solid particles of other shapes, however, is more difficult to analyse.

The force or drag F on a sphere is usually expressed in terms of a drag coefficient c_d defined by

$$c_d = \frac{F}{\frac{1}{2}\rho u^2 \cdot \frac{1}{4}\pi D^2}, \qquad (20.3.1)$$

where F is the force acting on the sphere, u is the velocity of the fluid relative to the sphere and D is the diameter of the sphere. The drag coefficient is a function of the Reynolds number as indicated for instance in fig. 75.

In the viscous range, with values of Re less than 2, Stokes's law applies, i.e.

$$c_d = \frac{24}{Re}.$$

For larger values of Re the experimental curve for the drag coefficient diverges from the Stokes law relationship. The following empirical formula may be used in the range from $Re = 2$ to $Re = 200$:

$$c_d = 18 \cdot 0 Re^{-0 \cdot 6}.$$

The curve of c_d subsequently flattens out to a value $c_d = 0\cdot44$ at $Re = 2000$, and it varies very little from this value in the range of Reynolds number from 2000 to about 200,000. At a greater value of Re, somewhere between 10^5 and 10^6, there is a sharp drop in the curve due to transition to turbulence in the boundary layer prior to separation. This, however, is well outside the usual range of Reynolds number appropriate to solid particles in fluid streams.

The *terminal settling velocity* of a spherical particle under gravity is given by

$$\frac{\pi}{6} D_p^3(\rho_s - \rho)g = c_d\tfrac{1}{2}\rho v_g^2 \frac{\pi D_p^2}{4},$$

where ρ is the density of the fluid, ρ_s is the density of the solid particle, D_p is the diameter of the particle and v_g is the terminal velocity, and hence

$$v_g = \left[\frac{4}{3}\frac{D_p(\rho_s - \rho)g}{c_d\rho}\right]^{\frac{1}{2}}. \qquad (20.3.2)$$

In the Stokes law range, for values of Re less than about 2,

$$v_g = \frac{D_p^2}{18}\frac{(\rho_s - \rho)g}{\mu}. \qquad (20.3.3)$$

The maximum particle diameter for Stokes's law to apply in the case of gravity settling can be found as follows:

$$D_p = \frac{3}{4}\frac{c_d v_g^2\rho}{(\rho_s - \rho)g} = \frac{18}{Re}v_g^2\frac{\rho}{(\rho_s - \rho)g},$$

i.e.
$$D_p^3 = \frac{18Re\,\mu^2}{\rho(\rho_s - \rho)g},$$

and putting $Re = 2$ for the critical case

$$D_{p_{\text{max.}}} = \left[\frac{36\mu^2}{\rho(\rho_s - \rho)g}\right]^{\frac{1}{3}}. \qquad (20.3.4)$$

These results may be applied to non-spherical particles using D_s, the diameter of the equivalent-volume sphere, in place of D_p and with a correction factor applied to the drag coefficient. The greater the departure from a spherical shape, however, the less will be the value of this method of analysis.

For *heat transfer* by conduction between a spherical particle and an infinite surrounding fluid, the following expression can be derived:

$$Nu = \frac{hD_p}{k} = 2.$$

This result will also apply with laminar flow past a sphere at very small values of the Reynolds number uD_p/ν. In terms of a Stanton number $h/\rho u c_p$ it takes the alternative form

$$St = \frac{2}{Re\ Pr}.$$

With turbulent flow, however, an empirical expression must be used for the heat-transfer coefficient, and the following form can be employed for values of the Reynolds number above about 20:

$$Nu = C Re^a Pr^b$$

or
$$St = C Re^{a-1} Pr^{b-1}.$$

The usual values assigned to the constants C, a and b are $C = 0.37$, $a = 0.6$ and $b = 0.3$.

The *suspension of solid particles in a turbulent fluid* can be investigated on the following lines. Let n be the number of particles per unit volume of fluid at any point. Assume that n is a function of the height y only, and that the particles are of uniform size and mass m_p. Due to turbulent mixing the rate of transfer of mass in a vertical direction will be

$$- m_p l' \tilde{v}' \frac{dn}{dy},$$

where l' is the mixing length and $\tilde{v}'$ is the r.m.s. value of the random fluctuating velocity component for turbulent motion of the particles.

If the terminal settling velocity of a particle is v_g the rate of transfer of mass downwards by the action of gravity will be

$$m_p n v_g.$$

Hence for steady conditions, if there is no net transfer of mass upwards or downwards,

$$n v_g = - l' \tilde{v}' \frac{dn}{dy}$$

or
$$\frac{dn}{n} = - \frac{v_g}{l' \tilde{v}'} dy. \tag{20.3.5}$$

If in the case of artificial mixing $l'\tilde{v}'$ can be assumed constant the resulting distribution of particle density will be

$$\log_e \frac{n}{n_0} = - \frac{v_g(y - y_0)}{l' \tilde{v}'}. \tag{20.3.6}$$

For turbulent flow in a horizontal pipe or channel, however, we can use Prandtl's mixing length hypothesis $l'\tilde{v}' = l^2(\mathrm{d}\bar{u}/\mathrm{d}y)$, and with a logarithmic velocity distribution for the fluid

$$\bar{u} = \frac{1}{k}\sqrt{\frac{\tau_0}{\rho}}\log_e y + \text{constant}$$

or

$$\frac{\mathrm{d}\bar{u}}{\mathrm{d}y} = \frac{1}{ky}\sqrt{\frac{\tau_0}{\rho}},$$

and therefore, if $\quad l = ky, \quad l'\tilde{v}' = ky\sqrt{\frac{\tau_0}{\rho}}.$

Hence from (20.3.5)

$$\frac{\mathrm{d}n}{n} = -\frac{v_g}{k\sqrt{(\tau_0/\rho)}}\frac{\mathrm{d}y}{y}.$$

and therefore

$$\log_e\frac{n}{n_0} = -\frac{v_g}{k\sqrt{(\tau_0/\rho)}}\log_e\frac{y}{y_0}$$

or

$$\frac{n}{n_0} = \left[\frac{y}{y_0}\right]^{-\frac{v_g}{k\sqrt{(\tau_0/\rho)}}}. \tag{20.3.7}$$

20.4. Separation of solid particles from fluids

The separation of solid matter from a fluid is a frequent operation in chemical engineering. Separation from a liquid or slurry is usually carried out by the method of *filtration*, and this operation will be discussed in Chapter 21 in connexion with the theory of flow through porous material.

Other methods of separation which are of industrial importance may be classified as follows:

Gravity separation is usually restricted to relatively large particle size. The ordinary settling tank represents the simplest form of gravity separator.

The principle of *inertial separation* is employed in centrifuges, impingement separators and cyclone dust collectors. The cyclone collector, which represents the most common method of extracting finely divided material from a gas, will be discussed in the next section.

Another method of removing solid material from a gas is to employ a liquid spray in a *scrubber*.

For the extraction of very finely divided solid particles from a gas an *electrostatic precipitator* may be used.

20.5. Cyclone collectors

The general arrangement of a cyclone dust collector is shown diagrammatically in fig. 107. Dust-laden gas enters tangentially and the solid particles are thrown radially outwards, hitting the cylindrical wall of the cyclone and dropping by gravity into the collecting cone.

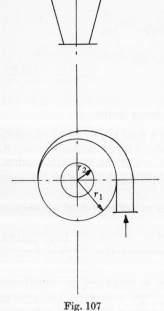

Fig. 107

Let the gas at radius r have velocity components u tangentially and v_r radially.

If a particle moves with the gas, with tangential velocity u, on an orbit of radius r, it will have a radial acceleration u^2/r. The ratio u^2/rg is sometimes called the 'separation factor'. If the particle keeps moving on the same orbit of radius r the radial acceleration u^2/r must be caused by an applied radial force exerted on the particle by the inward relative velocity of the gas. In the limiting case for equilibrium at radius r,

$$\frac{\pi}{6}(\rho_s - \rho)D_p^3 \frac{u^2}{r} = 3\pi D_p \mu v_r,$$

assuming that the particle diameter is small enough for Stokes's law to apply, i.e.

$$v_r = \frac{D_p^2}{18} \frac{(\rho_s - \rho)}{\mu} \frac{u^2}{r} \quad (20.5.1)$$

or

$$v_r = v_g \frac{u^2}{rg},$$

where v_g is the terminal settling velocity under gravity.

From (20.5.1) the particle diameter for equilibrium at radius r is given by

$$D_p^2 = \frac{18\mu}{(\rho_s - \rho)} \frac{rv_r}{u^2}. \quad (20.5.2)$$

For the radial velocity component v_r in the cylindrical barrel of the cyclone, we can say approximately $v_r = Q/2\pi r L$, where Q is the volume flow of gas and L is the effective length of the barrel, and hence

$$\frac{v_r}{v_{r1}} = \frac{r_1}{r}. \tag{20.5.3}$$

For the tangential velocity component u one might expect that the free vortex condition would apply, i.e. $ur = \text{constant} = u_1 r_1$. It is found experimentally, however, that, owing to the effects of skin friction at the wall, the tangential velocity component is given more nearly by a relation of the form

$$ur^{\frac{1}{2}} = u_1 r_1^{\frac{1}{2}} = \text{constant}. \tag{20.5.4}$$

Hence from (20.5.2) the particle diameter for equilibrium at radius r is given by

$$D_p^2 = \frac{18\mu}{(\rho_s - \rho)} \frac{v_{r1}}{u_1^2} r. \tag{20.5.5}$$

Putting $r = r_1$ in this expression, the critical particle diameter is given by

$$D_{p\text{crit.}}^2 = \frac{18\mu}{(\rho_s - \rho)} \frac{v_{r1} r_1}{u_1^2}. \tag{20.5.6}$$

Particles having a diameter larger than $D_{p\text{crit.}}$ will therefore reach the wall at $r = r_1$ and thus be trapped. Particles having diameters smaller than $D_{p\text{crit.}}$ will theoretically move on orbits of radius less than r_1, but in practice owing to velocity fluctuations they are liable to escape with the gas leaving the exit duct. It should be noted, however, that a cyclone collector does not achieve such a sharp cut as the simple theory would suggest.

The pressure drop for gas flow through a cyclone collector may be calculated as follows:

Inlet pressure drop $\Delta p_i = \frac{1}{2}\rho u_1^2.$

For the vortex $\dfrac{1}{\rho} \dfrac{\mathrm{d}p}{\mathrm{d}r} = \dfrac{u^2}{r} = \dfrac{u_1^2 r_1}{r^2},$

i.e. vortex pressure drop $\Delta p_v = \rho u_1^2 \left(\dfrac{r_1}{r_2} - 1 \right).$

Exit pressure drop $\Delta p_c = \frac{1}{2}\rho v_2^2,$

where v_2 is the velocity in the exit pipe.

The total pressure drop is therefore given by

$$\Delta p = \tfrac{1}{2}\rho u_1^2 + \rho u_1^2 \left(\frac{r_1}{r_2} - 1 \right) + \tfrac{1}{2}\rho v_2^2$$

or
$$\Delta p = \tfrac{1}{2}\rho u_1^2 \left(2\frac{r_1}{r_2} - 1 \right) + \tfrac{1}{2}\rho v_2^2. \qquad (20.5.7)$$

It is frequently found, however, that owing to rotation in the outlet pipe the exit pressure drop is approximately twice the theoretical value given above, in which case the last equation becomes

$$\Delta p = \tfrac{1}{2}\rho u_1^2 \left(2\frac{r_1}{r_2} - 1 \right) + \rho v_2^2. \qquad (20.5.7\,a)$$

all the pressure term also homogeneous. And we have, written in terms of r and the dimensionless groups, the equations to be solved. Thus, if we chose to perform the flow past a sphere problem in terms of a dimensionless property of the sphere radius . . . .

CHAPTER 21

FLOW THROUGH PACKED BEDS
AND FLUIDIZED SOLIDS

21.1. Viscous flow through a bed of solid particles

Consider a packed bed of solid particles. The *porosity* or *fraction void* is defined by

$$\varepsilon = \frac{\text{volume of voids}}{\text{total volume}} = \frac{A_e}{A},$$

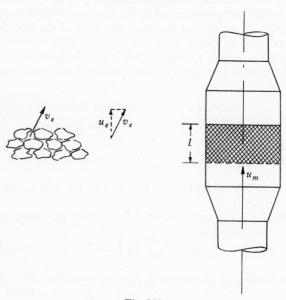

Fig. 108

where A is the total cross-sectional area of the bed and A_e is the effective cross-sectional flow area of the voids.

If a fluid is flowing through the bed at a total volumetric flow rate Q, the effective mean axial velocity component in the voids will be

$$u_e = \frac{Q}{A_e} = \frac{Q}{A}\frac{1}{\varepsilon} = \frac{u_m}{\varepsilon},$$

where u_m is the *mean approach velocity*. Owing to the tortuous nature of the paths through the bed, the effective length l_e of a channel will be greater than the height of the bed l. The effective mean value of the absolute velocity of the fluid passing through the spaces between the particles of the bed will therefore be given by

$$v_e = u_e \frac{l_e}{l} = \frac{u_m}{\varepsilon} \frac{l_e}{l}.$$

Referring back to the Hagen-Poiseuille law for viscous flow in pipes (equation (6.2.5)), we might expect the pressure drop for flow through a packed bed to be given by an expression of the form

$$\Delta p = \frac{k_0 \mu v_e l_e}{m^2}, \qquad (21.1.1)$$

where m is the hydraulic radius for the void passages, v_e is the mean velocity in the void passages and k_0 is a constant. Comparing (21.1.1) with (6.2.5) and noting that $4m$ takes the place of the pipe diameter D, the constant k_0 should have the value 2. This value, however, applies strictly only to passages of circular cross-section.

For the packed bed the hydraulic radius of the void passages is given by

$$m = \frac{\text{flow area}}{\text{wetted perimeter}} = \frac{\text{volume of fluid in the bed}}{\text{wetted surface}},$$

i.e. $\qquad m = \frac{\varepsilon A l}{\text{wetted surface}} = \frac{\varepsilon}{S},$

where S is the wetted surface of the particles per unit volume of the vessel.

Substituting for v_e and m in (21.1.1) we have the result

$$\Delta p = k_0 \frac{S^2}{\varepsilon^3} \left(\frac{l_e}{l}\right)^2 \mu u_m l, \qquad (21.1.2)$$

which is known as the Carman-Kozeny equation. This may be written

$$\Delta p = k \frac{S^2}{\varepsilon^3} \mu u_m l, \qquad (21.1.2a)$$

where $k = k_0(l_e/l)^2$. It is found experimentally that the value of the constant k is about 5·0 in many practical cases. This would imply that, if the average angle of the fluid passage through the bed is inclined at 45° to the axis so that $l_e/l = \sqrt{2}$, the constant k_0 in (21.1.1) would have the value 2·5 compared with the normal value 2·0 for a circular pipe.

For the special case of a packed bed of uniform solid particles having a diameter for the equivalent-volume sphere of D_s and a shape factor of λ,

number of particles per unit volume of vessel $= \dfrac{(1 - \varepsilon)}{\frac{1}{6}\pi D_s^3}$,

therefore $S = \lambda \pi D_s^2 \times$ number per unit volume $= \dfrac{6(1 - \varepsilon)\lambda}{D_s}$.

Hence from (21.1.2a), taking $k = 5\cdot 0$,

$$\Delta p = 180 \lambda^2 \frac{(1 - \varepsilon)^2}{\varepsilon^3} \frac{\mu u_m l}{D_s^2} \tag{21.1.3}$$

or $$\frac{\Delta p}{\rho u_m^2} = 180 \lambda^2 \frac{(1 - \varepsilon)^2}{\varepsilon^3} \frac{l}{D_s} \left(\frac{\nu}{u_m D_s} \right). \tag{21.1.3a}$$

Note carefully that these results only apply to fully viscous flow at small values of the Reynolds number.

For a bed of particles of *mixed sizes* but of similar shape:

total particle surface area $= \lambda \pi \Sigma D_s^2 = \lambda \pi N D_{sm}^2$,

where D_{sm} is the surface mean diameter (see § 20.1)

total volume of solid matter $= \tfrac{1}{6}\pi \Sigma D_s^3 = \tfrac{1}{6}\pi N D_{vm}^3$,

where D_{vm} is the volume mean diameter, i.e.

particle surface area per unit volume of solid matter

$$= 6\lambda \frac{\Sigma D_s^2}{\Sigma D_s^3} = 6\lambda \frac{D_{sm}^2}{D_{vm}^3} = \frac{6\lambda}{D_{vsm}},$$

where D_{vsm} is the *volume-surface mean diameter*. Hence the particle surface per unit volume of vessel $S = \dfrac{6(1 - \varepsilon)\lambda}{D_{vsm}}$, and the equation for the pressure drop becomes

$$\Delta p = 180\,\lambda^2 \frac{(1 - \varepsilon)^2}{\varepsilon^3} \frac{\mu u_m l}{D_{vsm}^2}, \tag{21.1.4}$$

i.e. the same form as (21.1.3) but with the volume-surface mean diameter D_{vsm} in place of D_s.

21.2. General analysis of flow through packed beds

We can proceed as before by analogy with flow through a pipe. Using a skin-friction coefficient c_f the pressure drop for flow in a pipe is given by (6.1.3), i.e.

$$\frac{\Delta p}{\rho u_m^2} = 2c_f \frac{l}{D},$$

where D is the pipe diameter, u_m is the mean velocity, and c_f is a function of $u_m D/\nu$.

For flow through the passages of a packed bed we could say

$$\frac{\Delta p}{\rho v_e^2} = 2c_f \frac{l_e}{4m}, \tag{21.2.1}$$

or

$$\frac{\Delta p}{\rho v_e^2} = c_f \frac{l_e}{2m},$$

where m is the hydraulic radius for the void passages, v_e is the mean effective velocity, and l_e is the effective length of a path through the bed. The coefficient c_f should be a function of a Reynolds number of the form $\left(\dfrac{4mv_e}{\nu}\right)$.

It was shown in §21.1 that

$$v_e = \frac{u_m}{\varepsilon} \frac{l_e}{l}$$

and

$$m = \frac{\varepsilon}{S} = \frac{\varepsilon}{(1-\varepsilon)} \frac{D_s}{6\lambda};$$

hence the logical Reynolds number is given by

$$\frac{4mv_e}{\nu} = \frac{2}{3} \frac{D_s}{\lambda} \frac{l_e}{l} \frac{u_m}{(1-\varepsilon)\nu}, \tag{21.2.2}$$

and from (21.2.1) the coefficient c_f is given by

$$c_f = \frac{2m}{l_e} \frac{\Delta p}{\rho v_e^2} = \frac{1}{3}\left(\frac{l}{l_e}\right)^3 \frac{\varepsilon^3}{\lambda(1-\varepsilon)} \frac{D_s}{l} \frac{\Delta p}{\rho u_m^2}. \tag{21.2.3}$$

c_f should then be a function of the logical Reynolds number as defined by (21.2.2).

The l/l_e term should remain at much the same value in all cases, and it may therefore be ignored in the final definition of a Reynolds number and a friction coefficient. The modified Reynolds number Re' and friction coefficient k_f are defined by

$$Re' = \frac{2}{3} \frac{1}{\lambda(1-\varepsilon)}\left(\frac{u_m D_s}{\nu}\right) \tag{21.2.4}$$

and

$$k_f = \frac{1}{3} \frac{\varepsilon^3}{\lambda(1-\varepsilon)} \frac{D_s}{l} \frac{\Delta p}{\rho u_m^2}. \tag{21.2.5}$$

Also from the definition of k_f the equivalent of the Fanning equation for the pressure drop Δp may be written

$$\frac{\Delta p}{\rho u_m^2} = 3k_f \frac{\lambda(1-\varepsilon)}{\varepsilon^3} \frac{l}{D_s}. \tag{21.2.6}$$

In the *viscous flow region* from the Carman-Kozeny equation (21.1.3)

$$3k_f\lambda\,\frac{(1-\varepsilon)}{\varepsilon^3}\,\frac{l}{D_s} = 180\lambda^2\,\frac{(1-\varepsilon)^2}{\varepsilon^3}\,\frac{l}{D_s}\left(\frac{\nu}{u_m D_s}\right)$$

or
$$k_f = \frac{40}{Re'}. \qquad (21.2.7)$$

The general form for k_f as a function of Re' is shown in fig. 109.

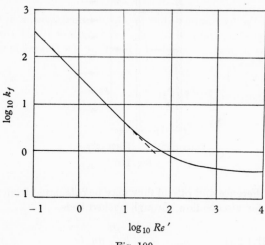

Fig. 109

21.3. Filtration

A great variety of filtration methods are used in the chemical industry for the purpose of separating solid material from liquids or slurries. The ordinary plate-and-frame filter press working on a batch system is perhaps the most common type of filtration equipment. Rotary filters of various specialized designs, however, are used in large-scale continuous processes. The essential mechanism in either case is to pump the liquid through a porous filter cloth so that the solid matter adheres to the cloth and builds up to form a 'cake'. The idealized process is shown diagrammatically in fig. 110 and may be analysed in the following way. Referring to the Carman-Kozeny equation (21.1.2*a*) we can say that for viscous flow through the filter cloth

$$p_1 - p_0 = c\mu u_m, \qquad (21.3.1)$$

where c is a constant which takes account of the thickness of the cloth and its effective porosity, etc., after it has been plugged with solid material.

18

The filter cake will build up in thickness during the period of filtration and there are two possible cases which must be considered regarding the behaviour of the cake and the nature of the flow:

(i) If the cake is incompressible, i.e. if the porosity is uniform throughout the thickness of the cake, the relationship between

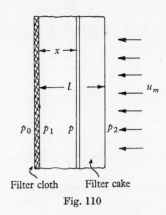

Filter cloth   Filter cake

Fig. 110

pressure difference and rate of flow may be calculated from the usual equations for viscous flow through packed beds.

From (21.1.2*a*)
$$p_2 - p_1 = k\frac{S^2}{\varepsilon^3}\mu u_m l$$

or
$$p_2 - p_1 = r\mu u_m l, \tag{21.3.2}$$

where $r = k(S^2/\varepsilon^3)$ is the 'specific resistance' of the cake per unit thickness. Hence from (21.3.1) and (21.3.2)

$$\Delta p = (p_2 - p_0) = \mu u_m(rl + c). \tag{21.3.3}$$

(ii) If the cake is compressible its porosity will vary with the 'solid pressure' p'. This is related to the fluid pressure p by

$$p' + p = p_2, \tag{21.3.4}$$

where p' and p are the solid pressure and the hydrostatic fluid pressure respectively at depth x in the cake, i.e. referring to fig. 110 p' rises from 0 at $x = l$ to the value $p_2 - p_1$ at $x = 0$.

In place of (21.3.2) we must now say

$$dp = r\mu u_m\, dx, \tag{21.3.5}$$

where r is the local specific resistance which will be a function of p'.

Noting from (21.3.4) that $dp' = -dp$, the last equation (21.3.5) may be written

$$-\frac{dp'}{r} = \mu u_m \, dx,$$

and hence

$$p_2 - p_1 = \bar{r}\mu u_m l, \qquad (21.3.6)$$

where $\bar{r}$ is the average specific resistance defined by

$$\frac{p_2 - p_1}{\bar{r}} = \int_0^{p_2 - p_1} \frac{dp'}{r}.$$

Hence from (21.3.1) and (21.3.6) the overall pressure difference is

$$\Delta p = (p_2 - p_0) = \mu u_m (\bar{r}l + c), \qquad (21.3.7)$$

which is similar to (21.3.3).

In either case the rate of accumulation of the cake is given by

$$\frac{dl}{dt} = \sigma u_m,$$

where σ is the volume fraction or concentration of solid matter in the fluid. Hence from (21.3.7), eliminating u_m,

$$\frac{dl}{dt} = \frac{\sigma \Delta p}{\mu(\bar{r}l + c)}. \qquad (21.3.8)$$

If the filtration is carried out with a *constant pressure difference*, equation (21.3.8) may be integrated to give

$$\mu\left(\frac{\bar{r}l^2}{2} + cl\right) = \sigma\Delta p t$$

or

$$t = \frac{\mu}{\sigma\Delta p}\left(\bar{r}\frac{l^2}{2} + cl\right). \qquad (21.3.9)$$

and since $l = \int \sigma u_m \, dt = \sigma V$, where V is the total volume of liquid passed through the filter per unit area, the last result may also be written

$$t = \frac{\mu}{\Delta p}\left(\bar{r}\frac{\sigma V^2}{2} + cV\right). \qquad (21.3.9a)$$

If, on the other hand, the filtration is carried out with a *constant volumetric flow rate* the time required to filter a volume V per unit area of filter is simply

$$t = \frac{V}{u_m},$$

and the pressure drop across the filter after time t is given by

$$\Delta p = \mu u_m (\bar{r}\sigma V + c)$$

or
$$\Delta p = \frac{\mu}{t}(\bar{r}\sigma V^2 + cV). \qquad (21.3.10)$$

The operation of a continuous rotary vacuum filter is illustrated diagrammatically in fig. 111. The filter cloth is stretched over the outer surface of the drum, and the interior of the drum is divided into a number of segments which are connected to the ports of a rotary valve. The drum revolves slowly in a tank of slurry and suction

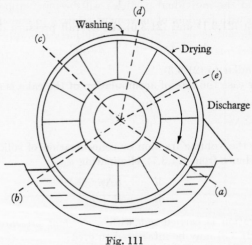

Fig. 111

is applied to the interior segments of the drum between positions (a) and (b) for the actual process of filtration. The filter cake builds up on the surface of the drum as it rotates through this region. The build-up of the cake between (a) and (b) may be calculated from equation (21.3.8) for constant pressure filtration. Beyond position (b) as the cake emerges from the tank of slurry it will at first be saturated with filtrate. It is usual to arrange for a wash to be applied by means of a spray in order to remove filtrate from the cake. Subsequently the cake may be dried by sucking air between positions (d) and (e). Finally a positive pressure is applied to the segment and the cake is removed from the drum by means of a scraper or other devices.

21.4. Fluidized solids

Starting with the case of flow of a gas through a static packed bed, with mean approach velocity u_m, as illustrated diagrammatically in fig. 108, consider what happens if the flow rate is steadily increased. The pressure difference Δp will be related to the flow velocity u_m by equation (21.2.6), i.e.

$$\Delta p = 3k_f \lambda \frac{(1 - \varepsilon)}{\varepsilon^3} \frac{l}{D_s} \rho u_m^2, \qquad (21.4.1)$$

where k_f is a function of the modified Reynolds number Re' defined by (21.2.4).

The pressure difference will increase with increasing rate of flow of the gas until, at a certain point, the bed of solid particles will expand slightly and the individual particles will become supported in the gas stream with freedom of movement relative to one another. The bed is then said to be *fluidized* and will have the appearance of a

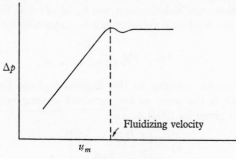

Fig. 112

boiling liquid with well-defined free surface and with bubbles of gas bursting at the surface. The relationship between pressure drop and rate of flow will be essentially as shown on fig. 112.

For a limited range of gas velocity, the fluidized bed remains in the so-called 'dense phase' appearing very much as a liquid and retaining its free surface. With uniform fluidization the gas flow should be well distributed and should emerge from the surface in quite small bubbles. Under certain conditions, however, either 'slugging' or 'channeling' of the gas flow may occur, and in these cases the flow is unsteady and the motion of the solid particles is of a much more violent nature. At higher gas velocities entrainment of the solid particles begins to take place and the bed expands into the 'dispersed phase' and loses its well-defined surface. The smaller particles are carried away first, but ultimately, if the velocity is increased sufficiently, all the particles will become entrained.

The critical velocity for the start of fluidization can be calculated as follows.

Under fluidized conditions the pressure difference must be related to the weight of solid particles supported in the gas stream by

$$\Delta p A = (\rho_s - \rho)g(1 - \varepsilon)Al$$

or
$$\Delta p = (1 - \varepsilon)(\rho_s - \rho)gl, \qquad (21.4.2)$$

where ε is the porosity at the point of fluidization, l is the height of the bed, ρ_s is the density of the solid material, ρ is the gas density. Up to the point of fluidization, however, the pressure difference is given by (21.4.1), hence eliminating Δp the *fluidizing velocity* is given by

$$u_m^2 = \frac{(\rho_s - \rho)}{\rho}\, g\, \frac{D_s \varepsilon^3}{3\lambda k_f}. \qquad (21.4.3)$$

If the flow through the bed of particles is viscous up to the point of fluidization, however, substituting for k_f in (21.4.3) from (21.2.7) and (21.2.4), the fluidizing velocity can be expressed by

$$u_m = (\rho_s - \rho)g\, \frac{D_s^2 \varepsilon^3}{180\mu\lambda^2(1 - \varepsilon)}. \qquad (21.4.4)$$

The entrainment velocity in the dispersed phase for a particle of diameter D_s is the same as the terminal settling velocity under gravity and is given by (20.3.2), i.e.

$$v_g = \left[\frac{4}{3}\, \frac{D_s(\rho_s - \rho)g}{c_d\, \rho}\right]^{\frac{1}{2}}.$$

where c_d is the drag coefficient for the particle. In designing a plant with a fluidized bed, therefore, the flow velocity in the bed must be fixed at some intermediate value between the fluidizing velocity and the entrainment velocity.

21.5. Heat and mass transfer in packed beds and fluidized solids

The problem of viscous flow through a packed bed with heat transfer between the solid particles and the fluid may be treated as one of longitudinal convection and transverse conduction with an effective thermal conductivity k_e whose value will depend both on the properties of the solid particles and of the fluid. The problem is therefore similar to that of forced convection with laminar flow in a circular pipe with a uniform velocity. The equation for the temperature distribution may be written

$$u_m \frac{\partial T}{\partial z} = \alpha_e \left[\frac{\partial^2 T}{\partial r^2} + \frac{1}{r}\frac{\partial T}{\partial r}\right], \qquad (21.5.1)$$

where
$$\alpha_e = \frac{k_e}{\rho c_p},$$

and similarly for mass transfer

$$u_m \frac{\partial c_A}{\partial z} = \mathscr{D}_e \left[\frac{\partial^2 c_A}{\partial r^2} + \frac{1}{r} \frac{\partial c_A}{\partial r} \right], \qquad (21.5.2)$$

where $\mathscr{D}_e$ is an effective diffusivity. The solution of these equations has already been discussed in §17.3.

The general problem of heat transfer under steady conditions between a fluid and the solid material of a packed bed, however, is best analysed on the lines of §21.2. We can define a special Stanton number for flow in the passages of the bed by $h/\rho v_e c_p$ and, noting that $v_e = \dfrac{u_m}{\varepsilon} \dfrac{l_e}{l}$, the special Stanton number must be related to the ordinary Stanton number defined in terms of the mean approach velocity u_m by

$$\frac{h}{\rho v_e c_p} = \frac{u_m}{v_e} St = \frac{l}{l_e} \varepsilon St,$$

where
$$St = \frac{h}{\rho u_m c_p}.$$

Also since
$$\frac{T_2 - T_1}{\Delta T_m} = \frac{l_e}{m} \frac{h}{\rho v_e c_p},$$

where T_1 and T_2 are the fluid inlet and outlet temperatures, ΔT_m is the mean temperature difference between solid and fluid, and m is the hydraulic radius for the void passages, the special Stanton number may be expressed alternatively by

$$\frac{h}{\rho v_e c_p} = \frac{l}{l_e} \frac{\varepsilon}{6\lambda(1 - \varepsilon)} \frac{D_s}{l} \frac{(T_2 - T_1)}{\Delta T_m}.$$

If we now omit the l/l_e term (as in §21.2 for the friction coefficient), we can define a modified Stanton number St' by

$$St' = \varepsilon \frac{h}{\rho u_m c_p} = \frac{1}{6\lambda} \frac{\varepsilon}{(1 - \varepsilon)} \frac{D_s}{l} \frac{(T_2 - T_1)}{\Delta T_m}, \qquad (21.5.3)$$

and this is analogous to the definition of k_f given by (21.2.5). As in the case of the friction coefficient, the modified Stanton number St' should be a function of the modified Reynolds number Re' defined by

$$Re' = \frac{2}{3} \frac{1}{\lambda(1 - \varepsilon)} \left(\frac{u_m D_s}{\nu} \right).$$

It is difficult, however, to determine this relationship from experimental results because most practical cases of heat transfer with a fluid flowing through a packed bed involve unsteady conditions, i.e. the solid material is heated up or cooled down over a period of time as in the case of a regenerator.

Two different aspects of heat transfer may arise with a *fluidized bed*. In some cases internal heating or cooling of the bed is carried out by means of tubes passing through the bed and carrying some fluid. In other cases, if a reaction is taking place, the heat of reaction is removed from or supplied to the bed by the gas stream itself. In the one case the problem is that of heat transfer between the gas/solid mixture and the surface of the tubes, while in the other case the problem is that of heat and mass transfer between the gas and the solid particles in the bed.

For an individual particle in a fluidized bed the heat transfer between particle and gas can be expressed by

$$Nu = \frac{hD_p}{k} = f\left(\frac{u_e D_p}{\nu}, Pr\right),$$

and this can usually be put in the form

$$Nu = C\left(\frac{u_e D_p}{\nu}\right)^a Pr^b, \qquad (21.5.4)$$

where C, a and b may be treated as constants for a reasonable range of Reynolds and Prandtl numbers.

The last result may be expressed alternatively by

$$St' = \frac{h}{\rho u_e c_p} = C\left(\frac{u_e D_p}{\nu}\right)^{a-1} Pr^{b-1}. \qquad (21.5.4a)$$

If experimental results are to be expressed in terms of a Reynolds number Re defined with the mean approach velocity u_m instead of u_e, and if a Stanton number St defined by $\dfrac{h}{\rho u_m c_p}$ is to be used in place of St', the last result must be written

$$St = \frac{h}{\rho u_m c_p} = C\varepsilon^{-a} Re^{a-1} Pr^{b-1}.$$

CHAPTER 22

CONDENSATION AND EVAPORATION

22.1. Film condensation

Condensation of a vapour in a surface condenser can take place in two different ways. The normal mechanism is that of *film condensation*, i.e. the condensed liquid forms a continuous film on the cooling surface and the vapour actually condenses on the liquid film. The other mechanism is that of *dropwise* condensation when individual drops are formed which detach themselves from the surface. Dropwise condensation can result in very high heat-transfer rates with values of h of 10,000 and upwards. However, it can only occur if the liquid does not wet the surface of the condenser and in practice it is very rarely achieved.

The mechanism of film condensation can be understood from the following analysis for condensation on a vertical surface, due to Nusselt.

It is assumed that the liquid film formed on the surface flows downwards with laminar motion and that the velocity distribution in the film is parabolic, i.e.

$$u = \frac{g}{\nu}\left(\delta y - \frac{y^2}{2}\right), \qquad (22.1.1)$$

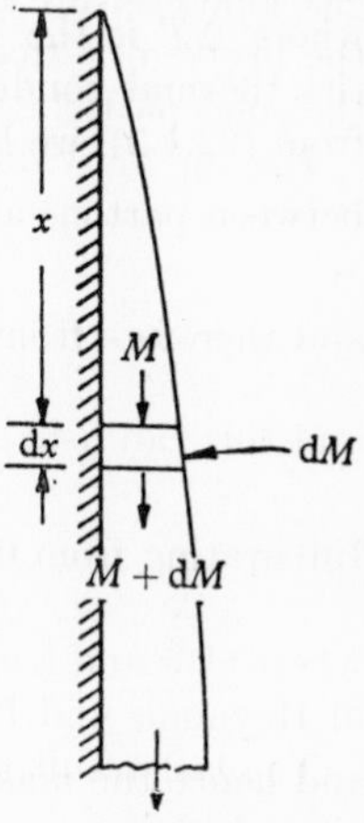

Fig. 113

where δ is the thickness of the film at distance x below the top of the surface.

The volume flow at any section under these conditions is given by

$$Q = \frac{g}{\nu}\frac{\delta^3}{3},$$

and the mass flow of liquid is therefore

$$M = \frac{\rho g}{\nu}\frac{\delta^3}{3}, \qquad (22.1.2)$$

i.e. the thickness of the film is related to the mass flow by

$$\delta = \left(\frac{3\nu M}{\rho g}\right)^{\frac{1}{3}}. \qquad (22.1.3)$$

265

With condensation taking place, the liquid film must increase both in mass flow and thickness as it flows down the surface. If dM is the rate of condensation (in lb./sec. or lb./hr.) in a length dx of the surface, a heat balance will give

$$L\,dM = q\,dx,\qquad\qquad(22.1.4)$$

where L is the latent heat, and q is the local heat flux. Assuming heat flow by conduction across the liquid film, with a linear transverse temperature distribution, we can say

$$q = k\frac{\Delta T}{\delta},$$

where ΔT is the local transverse temperature difference, and k is the thermal conductivity of the liquid. Hence, substituting for δ from (22.1.3), we have

$$q = k\Delta T\left(\frac{\rho g}{3\nu M}\right)^{\frac{1}{3}},\qquad\qquad(22.1.5)$$

and therefore from (22.1.4)

$$dM = \frac{k\Delta T}{L}\left(\frac{\rho g}{3\nu M}\right)^{\frac{1}{3}}dx.$$

Integrating from 0 to x, assuming that ΔT is constant,

$$\tfrac{3}{4}M^{\frac{4}{3}} = \frac{k\Delta T}{L}\left(\frac{\rho g}{3\nu}\right)^{\frac{1}{3}}x,$$

and hence the mass flow at distance x from the top of the surface is given by

$$M = \left[\frac{4}{3}\frac{k\Delta Tx}{L}\right]^{\frac{3}{4}}\left(\frac{\rho g}{3\nu}\right)^{\frac{1}{4}}.\qquad\qquad(22.1.6)$$

The total rate of heat flow is equal to ML, and therefore the average heat-transfer factor $h_{\text{av.}}$ is given by

$$h_{\text{av.}} = \frac{ML}{x\Delta T},$$

and the average value of the Nusselt number for a height x of the surface is

$$Nu = \frac{h_{\text{av.}}x}{k} = \frac{ML}{k\Delta T}.$$

Substituting for the mass flow M from (22.1.6)

$$Nu = (\tfrac{4}{3}x)^{\frac{3}{4}}\left(\frac{L}{k\Delta T}\right)^{\frac{1}{4}}\left(\frac{\rho g}{3\nu}\right)^{\frac{1}{4}}$$

or
$$Nu = 0.943 \left(\frac{L\rho g}{k\Delta T v}\right)^{\frac{1}{4}} x^{\frac{3}{4}}. \qquad (22.1.7)$$

This result may be expressed in an alternative form, in terms of the mass flow M per unit width of surface at distance x from the top, as follows:

the average value of the heat-transfer factor $h_{av.} = \dfrac{ML}{x\Delta T}$,

and from (22.1.6) $\quad k\Delta T x = \frac{3}{4} M^{\frac{4}{3}} L \left(\dfrac{3v}{\rho g}\right)^{\frac{1}{3}},$

therefore
$$\frac{h_{av.}}{k} = \frac{4}{3} M^{-\frac{1}{3}} \left(\frac{\rho g}{3v}\right)^{\frac{1}{3}}$$

$$= \left(\tfrac{4}{3}\right)^{\frac{4}{3}} \left(\frac{4M}{\mu}\right)^{-\frac{1}{3}} \left(\frac{g}{v^2}\right)^{\frac{1}{3}}$$

or
$$\frac{h_{av.}}{k} \left(\frac{v^2}{g}\right)^{\frac{1}{3}} = 1.47 \left(\frac{4M}{\mu}\right)^{-\frac{1}{3}}. \qquad (22.1.8)$$

Note that $(v^2/g)^{\frac{1}{3}}$ has dimensions of length L and the group on the left-hand side of the equation is therefore in the form of a Nusselt number.

The ratio $\dfrac{4M}{\mu} = \dfrac{4\rho u_m \delta}{\mu}$, where δ is the thickness of the liquid film at the bottom of the surface. This is the ordinary definition of a Reynolds number for flow in a channel with hydraulic mean depth δ and velocity u_m.

It is found experimentally that the flow in the liquid film is usually turbulent if the Reynolds number $4M/\mu$ is numerically greater than 2000. In these circumstances, according to [1], the heat-transfer rate is given by

$$\frac{h_{av.}}{k} \left(\frac{v^2}{g}\right)^{\frac{1}{3}} = 0.0077 \left(\frac{4M}{\mu}\right)^{0.4}. \qquad (22.1.9)$$

For further discussion of the mechanism of film condensation, including the effect of vapour velocity on condensation, the reader should consult [1].

22.2. Heat transfer to boiling liquids

There are three possible stages which must be considered in the transfer of heat from a flat metal surface to a boiling liquid:

(i) Transfer of heat to the liquid at a temperature below the local boiling-point. This will occur, for instance, over the lower portions of the submerged heating surface of an evaporator where boiling is suppressed locally owing to the hydrostatic head of the liquid above. For free convection to a liquid from horizontal or vertical heating surfaces in this manner the usual expressions for heat transfer given in §10.5 may be used.

(ii) The term 'nucleate boiling' is used to describe the process in those regions of an evaporator where small bubbles of vapour form on the heating surface and subsequently detach themselves and rise through the liquid. With nucleate boiling the heat-transfer factor h varies with the temperature difference ΔT approximately in the following manner:

$$h = \frac{q}{\Delta T} = \text{constant} \times \Delta T^{2\cdot 5}, \qquad (22.2.1)$$

and the heat flux q is therefore proportional to $\Delta T^{3\cdot 5}$.

(iii) If the temperature of the heating surface is sufficiently high, a continuous film of vapour may be formed on the surface, and the

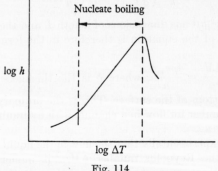

Fig. 114

heat-transfer factor will then fall sharply from the value that it would have under conditions of nucleate boiling.

The three zones of single-phase heating, nucleate boiling and film boiling are shown for a submerged heating surface on fig. 114, which gives the general form of the curve of heat-transfer factor plotted against the temperature difference ΔT between heating surface and liquid. There will generally be a definite maximum value for the heat flux occurring at a certain critical temperature difference for heat transfer between the surface and the boiling liquid. This critical temperature difference depends on the physical properties of the fluid and also to some extent on the geometry of the heating surface. The maximum heat flux occurring at the critical temperature difference is sometimes described as the 'burn-out flux', since any increase of temperature of the heating surface beyond this point will actually result in a decrease in the rate of heat transfer.

The problem of heat transfer to a boiling liquid flowing upwards *inside a tube* differs in certain features from the case of heat transfer with flat surfaces. The problem is illustrated diagrammatically in fig. 115. The liquid is assumed to enter the bottom of the tube at some temperature below the boiling-point, and in the lower part of the tube the heat-transfer rate will be given by the ordinary expressions for forced convection such as

$$Nu = 0 \cdot 023 Re^{0 \cdot 8} Pr^{0 \cdot 4} \qquad (22.2.2)$$

for the case of turbulent flow.

If the tube-wall temperature is above the local boiling-point, however, local nucleate boiling may occur in the boundary-layer region adjacent to the wall. Bubbles of vapour formed within the thermal boundary layer will move outwards into the main stream of colder liquid and will then condense again. Actually the liquid in the thermal boundary layer must be slightly superheated, since, owing to the effect of curvature and surface tension, the vapour pressure inside a very small bubble must be greater than the pressure of the surrounding liquid.

With local nucleate boiling at the tube wall or in the thermal boundary layer, heat-transfer rates will tend to be greater than those normally obtained with single-phase heating, since the ordinary turbulent convection process is supplemented by the additional transfer mechanism of local boiling and condensation. Experimental results with this type of flow, however, are rather limited.

Referring again to fig. 115, since the local hydrostatic pressure in the tube decreases as the fluid moves up the tube, owing to the change of head with distance from the bottom of the tube, the local satura-tion temperature or boiling-point will also decrease. This is shown by the dotted curve in fig. 115. At a certain point in the flow, the bulk temperature of the fluid will reach the local boiling-point. This is shown by the intersection of the curve for the mean fluid tempera-ture with the dotted curve for the local saturation temperature. From this point onwards bulk boiling will occur within the body of the fluid, and the temperature of the liquid/vapour mixture will remain essentially at the local saturation value.

The orderly nature of the flow indicated in fig. 115, however, is not always observed in practice. Instead of nucleate bulk boiling in the upper portion of the tube, a continuous film of vapour may be formed at the tube wall. Large bubbles or pockets of vapour may also extend across the entire section of the tube, and these will flow upwards and emerge from the outlet of the tube with slugs of liquid being ejected intermittently. Generally in the case of flow of a liquid/vapour mixture the vapour will travel up the tube faster than the liquid, and one of the main difficulties in estimating the circulation rate in an evaporator is the determination of this relative slip velocity.

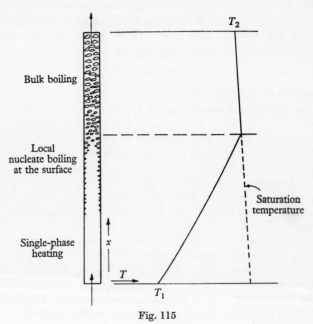

Fig. 115

22.3. Condensers and evaporators

While it is of interest and importance to investigate the fundamental nature of the flow and transfer processes occurring in condensation and evaporation, these processes are so complex that it is usually necessary to resort to the use of overall coefficients when designing industrial condensers and evaporators. The appropriate numerical values must be determined from experiments with other equipment of a similar nature. Frequently the two processes of condensation and evaporation occur together in the same piece of plant, as, for example, when condensing steam or some other organic vapour is used as the heating agent for an evaporator in a chemical plant. Owing to the accumulation of scale and dirt, heat-transfer rates obtained in practice on an industrial scale are usually much lower than those obtained with clean surfaces under laboratory conditions.

Surface condensers are usually designed with horizontal tubes and with cooling water flowing inside the tubes. For condensation on the outside of a single horizontal tube an expression can be derived in a similar manner to the Nusselt formula for film condensation on a vertical surface

$$Nu = \frac{hd}{k} = 0.725 \left(\frac{L\rho g}{k\Delta T\nu}\right)^{\frac{1}{4}} d^{\frac{3}{4}},$$

and for a vertical tier of N horizontal tubes

$$Nu = 0.725 \left(\frac{L\rho g}{Nk\Delta T\nu}\right)^{\frac{1}{4}} d^{\frac{3}{4}},$$

where d is the outside diameter of the tubes.

The accumulation of any non-condensable gas in the vapour space inside the shell of a surface condenser will have an adverse effect on the heat-transfer rate. Although precautions are taken to remove non-condensable gases the arrangements for purging are seldom completely effective. For this and other reasons, it is frequently found that there is a considerable divergence between the heat-transfer rates observed in practice and the values calculated from the theory of film condensation.

Evaporators vary widely in their design features. They can be classified in various ways, e.g. boiling inside or outside tubes, vertical or horizontal tubes, and natural or forced circulation. For further information on heat-transfer coefficients the reader should consult [2] and [3].

REFERENCES

(1) COLBURN. Problems in design and research on condensers of vapours and vapour mixtures. *Proc. Instn Mech. Engrs*, **164**.
(2) McADAMS. *Heat Transmission* (McGraw-Hill).
(3) Proceedings of the general discussion on heat transfer (Institution of Mechanical Engineers, 1951).

APPENDIX 1

ENGINEER'S GUIDE TO VECTOR ANALYSIS

A *vector* quantity can be represented by

$$\mathbf{a} = a_x\mathbf{i} + a_y\mathbf{j} + a_z\mathbf{k},$$

where $\mathbf{i}$, $\mathbf{j}$, $\mathbf{k}$ are the unit vectors in the x, y and z directions, and a_x, a_y and a_z are the scalar components of the vector $\mathbf{a}$.

Scalar and vector products are defined as follows:

Scalar product $\qquad \mathbf{a} . \mathbf{b} = a_x b_x + a_y b_y + a_z b_z.$

Vector product

$$\mathbf{a} \times \mathbf{b} = \begin{vmatrix} \mathbf{i} & \mathbf{j} & \mathbf{k} \\ a_x & a_y & a_z \\ b_x & b_y & b_z \end{vmatrix}.$$

The vector operator 'nabla' or 'del' is defined by

$$\nabla = \mathbf{i}\frac{\partial}{\partial x} + \mathbf{j}\frac{\partial}{\partial y} + \mathbf{k}\frac{\partial}{\partial z}.$$

The *gradient of a scalar quantity* ϕ is then given by

$$\text{grad } \phi = \nabla\phi = \mathbf{i}\frac{\partial\phi}{\partial x} + \mathbf{j}\frac{\partial\phi}{\partial y} + \mathbf{k}\frac{\partial\phi}{\partial z}.$$

The *divergence* and *curl* of a vector quantity $\mathbf{a}$ are given by

$$\text{div } \mathbf{a} = \nabla . \mathbf{a} = \frac{\partial a_x}{\partial x} + \frac{\partial a_y}{\partial y} + \frac{\partial a_z}{\partial z}$$

$$\text{curl } \mathbf{a} = \nabla \times \mathbf{a} = \begin{vmatrix} \mathbf{i} & \mathbf{j} & \mathbf{k} \\ \dfrac{\partial}{\partial x} & \dfrac{\partial}{\partial y} & \dfrac{\partial}{\partial z} \\ a_x & a_y & a_z \end{vmatrix}.$$

Note that div $\mathbf{a}$ is a scalar quantity, but that grad ϕ and curl $\mathbf{a}$ are both vector quantities.

The divergence and curl of a product $(\phi\mathbf{a})$ are sometimes required:

$$\text{div } (\phi\mathbf{a}) = (\text{grad } \phi) . \mathbf{a} + \phi \text{ div } \mathbf{a},$$

$$\text{curl } (\phi\mathbf{a}) = (\text{grad } \phi) \times \mathbf{a} + \phi \text{ curl } \mathbf{a}.$$

The divergence of grad ϕ is an important function and can be expressed by

$$\text{div } (\text{grad } \phi) = \nabla . \nabla\phi = \nabla^2\phi = \frac{\partial^2\phi}{\partial x^2} + \frac{\partial^2\phi}{\partial y^2} + \frac{\partial^2\phi}{\partial z^2}.$$

The operator ∇^2 can also be applied to a vector quantity **a** with the meaning

$$\nabla^2 \mathbf{a} = \frac{\partial^2 \mathbf{a}}{\partial x^2} + \frac{\partial^2 \mathbf{a}}{\partial y^2} + \frac{\partial^2 \mathbf{a}}{\partial z^2}.$$

Other second-order differential vector quantities can be derived as follows:

$$\operatorname{div} \operatorname{curl} \mathbf{a} = \nabla \cdot \nabla \times \mathbf{a} \quad = 0,$$
$$\operatorname{curl} \operatorname{grad} \phi = \nabla \times \nabla \phi \quad = 0,$$
$$\operatorname{curl} \operatorname{curl} \mathbf{a} = \nabla \times \nabla \times \mathbf{a} = \operatorname{grad} \operatorname{div} \mathbf{a} - \nabla^2 \mathbf{a},$$
$$\operatorname{grad} \operatorname{div} \mathbf{a} = \nabla(\nabla \cdot \mathbf{a}) \quad = \operatorname{curl} \operatorname{curl} \mathbf{a} + \nabla^2 \mathbf{a}.$$

Gauss's theorem states that, if **a** is a continuous vector function of the position vector **r**, then considering a surface S enclosing a volume V,

$$\oint_S \mathbf{a} \cdot \mathbf{n} \, dS = \oint_V (\operatorname{div} \mathbf{a}) \, dV,$$

i.e.   surface integral of **a** = volume integral of div **a**,

n being the unit normal vector at any point on the surface.

Stokes's theorem states that, if **a** is a continuous vector function of the position vector **r**, then considering a closed curve L forming the boundary of a surface S,

$$\oint_L \mathbf{a} \cdot d\mathbf{L} = \oint_S (\operatorname{curl} \mathbf{a}) \cdot \mathbf{n} \, dS,$$

i.e.   the line integral of **a** = surface integral of curl **a**.

In fluid mechanics the velocity vector is represented by

$$\mathbf{u} = u\mathbf{i} + v\mathbf{j} + w\mathbf{k},$$

and this is expressed as a function of time t and of the position vector

$$\mathbf{r} = x\mathbf{i} + y\mathbf{j} + z\mathbf{k},$$

i.e. $\mathbf{u} = f(\mathbf{r}, t).$

The *total differential*

$$\frac{D\mathbf{u}}{Dt} = \frac{\partial \mathbf{u}}{\partial t} + (\mathbf{u} \cdot \nabla) \mathbf{u}.$$

The *equation of continuity*, which is derived with the help of Gauss's theorem (§2.2), states that

$$\frac{\partial \rho}{\partial t} + \operatorname{div}(\rho \mathbf{u}) = 0$$

or $$\frac{\partial \rho}{\partial t} + \mathbf{u} \cdot \nabla \rho + \rho \operatorname{div} \mathbf{u} = 0,$$

which may also be written

$$\frac{D\rho}{Dt} + \rho \operatorname{div} \mathbf{u} = 0.$$

For steady incompressible flow with the density ρ constant the equation reduces to

$$\operatorname{div} \mathbf{u} = 0.$$

APPENDIX 2

EQUATIONS OF MOTION FOR AN INVISCID FLUID

Considering a volume V of fluid bounded by a control surface S, the equation of motion may be written

$$\oint_V \rho \frac{\mathrm{D}\mathbf{u}}{\mathrm{D}t} \, \mathrm{d}V = \oint_V \rho \mathbf{F} \, \mathrm{d}V - \oint_S p\mathbf{n} \, \mathrm{d}S,$$

where $\mathbf{F}$ is the body force (e.g. gravity) acting on the fluid per unit mass, and p is the pressure in the fluid.

By a theorem analogous to Gauss's theorem the surface integral of the pressure p can be equated to the volume integral of grad p, i.e.

$$\oint_S p\mathbf{n} \, \mathrm{d}S = \oint_V (\mathrm{grad}\ p) \, \mathrm{d}V,$$

and hence the equation of motion becomes

$$\rho \frac{\mathrm{D}\mathbf{u}}{\mathrm{D}t} = \rho \mathbf{F} - \mathrm{grad}\ p,$$

and this is known as *Euler's equation* for an inviscid fluid.

If $\mathbf{F}$ can be expressed as the gradient of a scalar potential $-\Omega$, and if we consider the case of *steady flow* of the fluid the equation becomes

$$(\mathbf{u} \cdot \nabla)\mathbf{u} = -\mathrm{grad}\ \Omega - \frac{1}{\rho}\ \mathrm{grad}\ p,$$

and this may be integrated following the motion of the fluid, i.e. *along a streamline*, giving the result

$$\int \frac{\mathrm{d}p}{\rho} + \frac{\mathbf{u}^2}{2} + \Omega = \mathrm{constant}.$$

If $\mathbf{F}$ is the force due to gravity, $\Omega = gz$, and therefore

$$\int \frac{\mathrm{d}p}{\rho} + \tfrac{1}{2}\mathbf{u}^2 + gz = \mathrm{constant},$$

which is the *Bernoulli equation* for flow along a streamline. The constant applies only to the streamline considered.

It can be shown that $(\mathbf{u} \cdot \nabla)\mathbf{u} = \tfrac{1}{2}\nabla(\mathbf{u}^2) - \mathbf{u} \times \mathrm{curl}\ \mathbf{u}$. The Euler equation for steady flow may therefore be written

$$\tfrac{1}{2}\nabla(\mathbf{u}^2) - \mathbf{u} \times \mathrm{curl}\ \mathbf{u} = -\mathrm{grad}\ \Omega - \frac{1}{\rho}\ \mathrm{grad}\ p$$

or

$$\frac{1}{\rho}\nabla p + \tfrac{1}{2}\nabla(\mathbf{u}^2) + \nabla\Omega = \mathbf{u} \times \mathrm{curl}\ \mathbf{u}.$$

In the special case when curl **u** is zero everywhere, this reduces to

$$\frac{1}{\rho}\nabla p + \tfrac{1}{2}\nabla(\mathbf{u}^2) + \nabla\Omega = 0,$$

and for an incompressible fluid with ρ constant, the last equation may be written

$$\nabla\left(\frac{p}{\rho} + \tfrac{1}{2}\mathbf{u}^2 + \Omega\right) = 0,$$

and hence

$$\frac{p}{\rho} + \tfrac{1}{2}\mathbf{u}^2 + \Omega = \text{constant},$$

and the constant now applies throughout the fluid.

Note that this result is only valid if curl **u** is zero. This is the condition for *irrotational flow*.

It can also be shown that if the flow is irrotational, i.e. if curl **u** is zero, the velocity vector **u** may be expressed as the gradient of a scalar potential

$$\mathbf{u} = -\operatorname{grad}\phi,$$

where ϕ is the *velocity potential*.

The equation of continuity then takes the form (for incompressible flow)

$$\operatorname{div}\operatorname{grad}\phi = 0$$

or

$$\nabla^2\phi = 0 \text{ which is } \textit{Laplace's equation.}$$

The mathematical treatment of irrotational flow consists in finding solutions to Laplace's equation which will satisfy certain specified boundary conditions. This will give the velocity distribution for the flow. The pressure distribution can then be derived from the Euler equation of motion.

If we are dealing with *two-dimensional irrotational flow*, the velocity components may be expressed in terms of the velocity potential ϕ by

$$u = -\frac{\partial\phi}{\partial x} \quad\text{and}\quad v = -\frac{\partial\phi}{\partial y}.$$

It is convenient also to introduce a *stream function* ψ defined by

$$u = -\frac{\partial\psi}{\partial y} \quad\text{and}\quad v = \frac{\partial\psi}{\partial x}.$$

The condition $\psi = \text{constant}$ then represents a streamline.

Streamlines and lines of constant velocity potential for two-dimensional irrotational flow will form an orthogonal curvilinear system. The conjugate complex quantity $(\phi + i\psi)$ may then be expressed as a function of the complex variable $(x + iy)$. This fact enables the analysis of two-dimensional flow of an ideal inviscid fluid to be carried out with the help of the special mathematical technique of complex numbers and functions of a complex variable.

APPENDIX 3

VORTICITY AND CIRCULATION

The *vorticity* in a fluid is a vector quantity having the same nature as angular velocity. It is defined by

$$\text{vorticity } \mathbf{w} = \text{curl } \mathbf{u} = \nabla \times \mathbf{u},$$

and this may be expressed as

$$\text{curl } \mathbf{u} = \begin{vmatrix} \mathbf{i} & \mathbf{j} & \mathbf{k} \\ \dfrac{\partial}{\partial x} & \dfrac{\partial}{\partial y} & \dfrac{\partial}{\partial z} \\ u & v & w \end{vmatrix},$$

or

$$\text{curl } \mathbf{u} = \xi\mathbf{i} + \eta\mathbf{j} + \zeta\mathbf{k}.$$

where the three components are specified by

$$\xi = \left(\frac{\partial w}{\partial y} - \frac{\partial v}{\partial z}\right), \quad \eta = \left(\frac{\partial u}{\partial z} - \frac{\partial w}{\partial x}\right), \quad \zeta = \left(\frac{\partial v}{\partial x} - \frac{\partial u}{\partial y}\right).$$

The physical significance of vorticity is best understood by imagining a small spherical element of fluid to be suddenly frozen. If the resulting solid element has rotation then the fluid has vorticity at the point considered. The numerical magnitude of the vorticity would be equal to twice the angular velocity of the solid sphere.

A *vortex line* is defined as a line drawn in the fluid so that the tangent at any point has the same direction as the vorticity vector at that point. Vortex lines drawn through every point on a small closed curve would form a *vortex tube*. This is analogous to the case of streamlines and stream tubes. For a vortex tube of small cross-section it may be shown that the product of the magnitude of the vorticity and the cross-sectional area of the tube must have the same value all along the tube. This constant product is known as the strength of the vortex. Vortex lines cannot begin or end in the middle of the fluid. They must either form closed curves or else they must begin or end on the boundaries of the fluid.

Circulation is associated with a closed circuit drawn in the fluid. It is defined as the line integral of the tangential velocity component taken once round the closed circuit, i.e.

$$\text{circulation } K = \oint_L \mathbf{u} \cdot d\mathbf{L},$$

It follows from Stokes's theorem (Appendix 1) that

$$\oint_L \mathbf{u} \cdot d\mathbf{L} = \oint_S (\text{curl } \mathbf{u}) \cdot \mathbf{n}\, dS = \oint_S \mathbf{w} \cdot \mathbf{n}\, dS,$$

i.e. the circulation round a closed curve is equal to the surface integral of the vorticity taken over any surface enclosed by the curve.

APPENDIX 4

STRESS COMPONENTS IN A VISCOUS FLUID AND THE EQUATIONS OF MOTION

It is necessary, first of all, to analyse the *rate of strain* of a fluid element. If the velocity vector for the fluid is $\mathbf{u}$ at position $\mathbf{r}$, and $\mathbf{u} + \mathrm{d}\mathbf{u}$ at position $\mathbf{r} + \mathrm{d}\mathbf{r}$, the differential velocity must be given by

$$\mathrm{d}\mathbf{u} = \mathrm{d}\mathbf{r} \cdot \nabla\mathbf{u},$$

or, written out fully in terms of the components,

$$\mathrm{d}u = \frac{\partial u}{\partial x}\,\mathrm{d}x + \frac{\partial u}{\partial y}\,\mathrm{d}y + \frac{\partial u}{\partial z}\,\mathrm{d}z,$$

$$\mathrm{d}v = \frac{\partial v}{\partial x}\,\mathrm{d}x + \frac{\partial v}{\partial y}\,\mathrm{d}y + \frac{\partial v}{\partial z}\,\mathrm{d}z,$$

$$\mathrm{d}w = \frac{\partial w}{\partial x}\,\mathrm{d}x + \frac{\partial w}{\partial y}\,\mathrm{d}y + \frac{\partial w}{\partial z}\,\mathrm{d}z,$$

and these may be expressed alternatively in the following form:

$$\mathrm{d}u = \tfrac{1}{2}(e_{xx}\,\mathrm{d}x + e_{yx}\,\mathrm{d}y + e_{zx}\,\mathrm{d}z) + \tfrac{1}{2}(0 - \zeta\,\mathrm{d}y + \eta\,\mathrm{d}z),$$

$$\mathrm{d}v = \tfrac{1}{2}(e_{xy}\,\mathrm{d}x + e_{yy}\,\mathrm{d}y + e_{zy}\,\mathrm{d}z) + \tfrac{1}{2}(\zeta\,\mathrm{d}x + 0 - \xi\,\mathrm{d}z),$$

$$\mathrm{d}w = \tfrac{1}{2}(e_{xz}\,\mathrm{d}x + e_{yz}\,\mathrm{d}y + e_{zz}\,\mathrm{d}z) + \tfrac{1}{2}(- \eta\,\mathrm{d}x + \xi\,\mathrm{d}y + 0),$$

where e_{xx}, e_{xy}, etc., are the *rate of strain components* defined by

$$e_{xx} = 2\frac{\partial u}{\partial x}, \quad e_{yy} = 2\frac{\partial v}{\partial y}, \quad e_{zz} = 2\frac{\partial w}{\partial z},$$

$$e_{xy} = e_{yx} = \left(\frac{\partial v}{\partial x} + \frac{\partial u}{\partial y}\right),$$

$$e_{yz} = e_{zy} = \left(\frac{\partial w}{\partial y} + \frac{\partial v}{\partial z}\right),$$

$$e_{zx} = e_{xz} = \left(\frac{\partial u}{\partial z} + \frac{\partial w}{\partial x}\right),$$

and ξ, η, ζ are the components of the vorticity curl $\mathbf{u}$ as given in Appendix 3.

The significance of these different quantities will be understood from fig. 116, which illustrates the translation, deformation and rotation of a fluid element in the two-dimensional case.

The three terms e_{xx}, e_{yy}, e_{zz} represent expansion or compression of the fluid as shown for the two-dimensional case in fig. 116 (a). Note that the sum of these three terms

$$e_{xx} + e_{yy} + e_{zz} = 2\left(\frac{\partial u}{\partial x} + \frac{\partial v}{\partial y} + \frac{\partial w}{\partial z}\right) = 2 \text{ div } \mathbf{u},$$

which must be zero for an incompressible fluid.

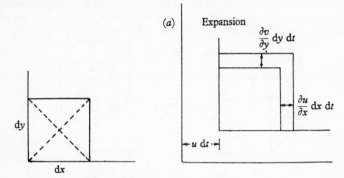

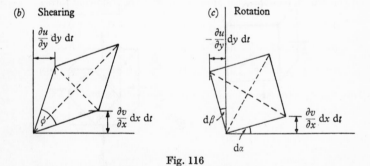

Fig. 116

The other six components of the rate of strain, e_{xy}, e_{yz}, etc., represent shearing or distortion of the fluid element. This is shown for the two-dimensional case in fig. 116 (b), where the rate of angular deformation $-\,\mathrm{d}\phi/\mathrm{d}t$ is given by $(\partial v/\partial x + \partial u/\partial y)$.

It will also be evident from fig. 116 (c) that the mean rate of angular rotation of the fluid element is given by

$$\frac{1}{2}\left(\frac{\partial \alpha}{\partial t} + \frac{\partial \beta}{\partial t}\right) = \frac{1}{2}\left(\frac{\partial v}{\partial x} - \frac{\partial u}{\partial y}\right) = \tfrac{1}{2}\,\zeta,$$

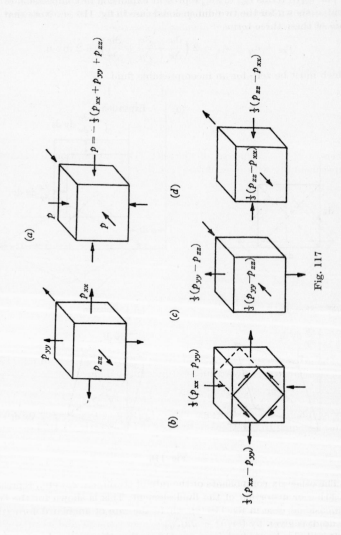

Fig. 117

and similarly for the other components of rotation, i.e. each vorticity component is equal to twice the corresponding component of angular velocity as stated in Appendix 3.

The analysis of the stresses in a viscous fluid has been outlined in §12.1. For a Newtonian fluid it is assumed that the *shear stresses* are proportional to the *rates of angular deformation*,

$$\tau_{xy} = \tau_{yx} = \mu \left(\frac{\partial v}{\partial x} + \frac{\partial u}{\partial y} \right),$$

$$\tau_{yz} = \tau_{zy} = \mu \left(\frac{\partial w}{\partial y} + \frac{\partial v}{\partial z} \right),$$

$$\tau_{zx} = \tau_{xz} = \mu \left(\frac{\partial u}{\partial z} + \frac{\partial w}{\partial x} \right),$$

where μ is the coefficient of viscosity. This assumption is confirmed experimentally with most fluids.

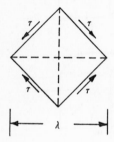
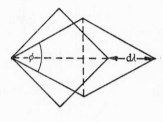

Fig. 118

The *direct stresses* can be resolved as shown in fig. 117. In fig. 117 (a) the element experiences a uniform pressure p defined by

$$p = - \tfrac{1}{3}(p_{xx} + p_{yy} + p_{zz}).$$

By symmetry the three linear rates of strain caused by the pressure p must be equal to one-third of the rate of volumetric strain, i.e.

$$\left(\frac{\partial u}{\partial x} \right)_a = - \frac{1}{3} \frac{1}{\rho} \frac{\mathrm{D}\rho}{\mathrm{D}t} = \tfrac{1}{3} \operatorname{div} \mathbf{u},$$

and similarly for the other components.

The stress system shown in fig. 117 (b) is equivalent to pure two-dimensional shearing of the inner element whose faces are inclined at 45° to those of the outer element. This inner element, and its subsequent state of distortion, is indicated in fig. 118. If λ is the length of the diagonal, the diagonal strain will be given by

$$\frac{\mathrm{d}\lambda}{\lambda} = - \tfrac{1}{2} \, \mathrm{d}\phi,$$

and the rate of strain

$$\left(\frac{\partial u}{\partial x}\right)_b = -\frac{1}{2}\frac{d\phi}{dt} = \frac{1}{2}\frac{\tau}{\mu}.$$

Hence, comparing fig. 117 (b) and fig. 118, since $\tau = \frac{1}{3}(p_{xx} - p_{yy})$

$$\left(\frac{\partial u}{\partial x}\right)_b = \frac{1}{2\mu}\frac{(p_{xx} - p_{yy})}{3}.$$

By similar reasoning from fig. 117 (d),

$$\left(\frac{\partial u}{\partial x}\right)_d = \frac{1}{2\mu}\frac{(p_{xx} - p_{zz})}{3},$$

and obviously $\qquad (\partial u/\partial x)_c = 0.$

Hence, adding the partial rates of strain,

$$\frac{\partial u}{\partial x} = \left(\frac{\partial u}{\partial x}\right)_a + \left(\frac{\partial u}{\partial x}\right)_b + \left(\frac{\partial u}{\partial x}\right)_c + \left(\frac{\partial u}{\partial x}\right)_d$$

$$= \tfrac{1}{3}\operatorname{div}\mathbf{u} + \frac{1}{2\mu}\frac{(p_{xx} - p_{yy})}{3} + \frac{1}{2\mu}\frac{(p_{xx} - p_{zz})}{3}$$

or $\qquad \dfrac{\partial u}{\partial x} = \tfrac{1}{3}\operatorname{div}\mathbf{u} + \dfrac{p_{xx}}{2\mu} - \dfrac{1}{2\mu}\dfrac{(p_{xx} + p_{yy} + p_{zz})}{3},$

i.e. $\qquad \dfrac{\partial u}{\partial x} = \tfrac{1}{3}\operatorname{div}\mathbf{u} + \dfrac{p_{xx}}{2\mu} + \dfrac{p}{2\mu},$

hence $\qquad p_{xx} = -p + 2\mu\,\dfrac{\partial u}{\partial x} - \tfrac{2}{3}\mu\operatorname{div}\mathbf{u},$

and similarly

$$p_{yy} = -p + 2\mu\,\frac{\partial v}{\partial y} - \tfrac{2}{3}\mu\operatorname{div}\mathbf{u}$$

and $\qquad p_{zz} = -p + 2\mu\,\dfrac{\partial w}{\partial z} - \tfrac{2}{3}\mu\operatorname{div}\mathbf{u}.$

These are the results quoted in (12.2.1).

Having derived the relationship between stress and rate of strain, the equations of motion follow directly by substituting for the stress components in (12.1.1). The resulting Navier-Stokes equation is given by (12.3.1), or written vectorially

$$\rho\,\frac{D\mathbf{u}}{Dt} = \rho\mathbf{F} - \operatorname{grad} p + \mu\nabla^2\mathbf{u} + \frac{\mu}{3}\operatorname{grad}\operatorname{div}\mathbf{u}.$$

LAMINAR BOUNDARY-LAYER FLOW

The equations for two-dimensional incompressible flow in a laminar boundary layer with zero pressure gradient are given by (13.1.3) and (13.1.4), i.e.

$$\frac{\partial u}{\partial x} + \frac{\partial v}{\partial y} = 0$$

and

$$u \frac{\partial u}{\partial x} + v \frac{\partial u}{\partial y} = \nu \frac{\partial^2 u}{\partial y^2}.$$

The equation of continuity can be satisfied by introducing a stream function ψ defined by

$$u = -\frac{\partial \psi}{\partial y} \quad \text{and} \quad v = \frac{\partial \psi}{\partial x}.$$

Substituting for u and v in the equation of motion, we thus have a partial differential equation for ψ in terms of the independent variables x and y, i.e.

$$\frac{\partial \psi}{\partial y} \frac{\partial^2 \psi}{\partial x \partial y} - \frac{\partial \psi}{\partial x} \frac{\partial^2 \psi}{\partial y^2} = -\nu \frac{\partial^3 \psi}{\partial y^3}.$$

However, by introducing the new variables η and f defined by

$$\eta = \frac{1}{2} \left(\frac{u_1}{\nu x} \right)^{\frac{1}{2}} y \quad \text{and} \quad f = \frac{-\psi}{(\nu u_1 x)^{\frac{1}{2}}}$$

the partial differential equation is reduced to an ordinary differential equation

$$\frac{\mathrm{d}^3 f}{\mathrm{d}\eta^3} + f \frac{\mathrm{d}^2 f}{\mathrm{d}\eta^2} = 0.$$

The physical meaning of this transformation is simply that the velocity profiles in the boundary layer are similar from section to section and that the boundary-layer thickness is proportional to $x^{\frac{1}{2}}$, where x is the distance measured from the leading edge. The new variable η is therefore proportional to the ratio y/δ.

The ordinary third-order differential equation may be solved as a series in terms of η. The solution satisfying the condition that $u = 0$ and $v = 0$ at the boundary where $\eta = 0$ is

$$f = \frac{\alpha \eta^2}{2!} - \frac{\alpha^2 \eta^5}{5!} + 11 \frac{\alpha^3 \eta^8}{8!} - \frac{375\ \alpha^4 \eta^{11}}{11!} + \dots.$$

α is a constant which has to be determined so that the remaining boundary condition, $u \to u_1$ as $y \to \infty$, is satisfied. The numerical value of

α which meets this requirement of the outer boundary condition is $\alpha = 1\cdot3282$. The velocity distribution as a function of η is then given by the following table:

η	u/u_1	η	u/u_1	η	u/u_1
0	0	1·0	0·630	2·0	0·956
0·1	0·066	1·1	0·681	2·1	0·967
0·2	0·133	1·2	0·729	2·2	0·976
0·3	0·199	1·3	0·773	2·3	0·983
0·4	0·265	1·4	0·812	2·4	0·988
0·5	0·330	1·5	0·846	2·5	0·992
0·6	0·394	1·6	0·876	2·6	0·994
0·7	0·456	1·7	0·902	2·7	0·996
0·8	0·517	1·8	0·923	2·8	0·997
0·9	0·575	1·9	0·941	2·9	0·998

The skin friction at the wall is given by

$$\tau_0 = \mu \left(\frac{\partial u}{\partial y} \right)_0 = \frac{\mu u_1}{4} \left(\frac{u_1}{vx} \right)^{\frac{1}{2}} \left(\frac{\mathrm{d}^2 f}{\mathrm{d}\eta^2} \right)_0 = \mu \frac{u_1 \alpha}{4} \left(\frac{u_1}{vx} \right)^{\frac{1}{2}},$$

and hence,

$$\frac{\tau_0}{\frac{1}{2}\rho u_1^2} = \frac{\alpha}{2} \left(\frac{u_1 x}{v} \right)^{-\frac{1}{2}} = 0\cdot664 \left(\frac{u_1 x}{v} \right)^{-\frac{1}{2}}.$$

The displacement thickness δ^* is given by

$$\frac{\delta^*}{x} = 1\cdot7208 \left(\frac{u_1 x}{v} \right)^{-\frac{1}{2}},$$

and the momentum thickness θ is given by

$$\frac{\theta}{x} = 0\cdot664 \left(\frac{u_1 x}{v} \right)^{-\frac{1}{2}}.$$

The drag coefficient for a length x is

$$c_D = 1\cdot328 \left(\frac{u_1 x}{v} \right)^{-\frac{1}{2}}.$$

APPENDIX 6

HEAT TRANSFER WITH LAMINAR
FLOW IN A PIPE

The energy equation for laminar rod-like flow in a circular pipe may be written as in (17.3.3)

$$u_m \frac{\partial \theta}{\partial z} = \alpha \left[\frac{\partial^2 \theta}{\partial r^2} + \frac{1}{r} \frac{\partial \theta}{\partial r} \right].$$

The method of solving a partial differential equation of this type is to try a solution of the form

$$\theta = R(r)Z(z),$$

where R is a function of r only, and Z is a function of z only. Then

$$\frac{\partial \theta}{\partial z} = R \frac{\mathrm{d}Z}{\mathrm{d}z}, \quad \frac{\partial \theta}{\partial r} = Z \frac{\mathrm{d}R}{\mathrm{d}r}, \quad \frac{\partial^2 \theta}{\partial r^2} = Z \frac{\mathrm{d}^2 R}{\mathrm{d}r^2},$$

and the equation becomes

$$\frac{u_m}{\alpha} R \frac{\mathrm{d}Z}{\mathrm{d}z} = Z \left[\frac{\mathrm{d}^2 R}{\mathrm{d}r^2} + \frac{1}{r} \frac{\mathrm{d}R}{\mathrm{d}r} \right]$$

or

$$\frac{u_m}{\alpha} \frac{1}{Z} \frac{\mathrm{d}Z}{\mathrm{d}z} = \frac{1}{R} \left[\frac{\mathrm{d}^2 R}{\mathrm{d}r^2} + \frac{1}{r} \frac{\mathrm{d}R}{\mathrm{d}r^2} \right].$$

The left-hand side is a function of z only and the right-hand side of r only. Suppose that each is equal to a constant $-\beta^2$. Then

$$\frac{1}{Z} \frac{\mathrm{d}Z}{\mathrm{d}z} = -\frac{\alpha}{u_m} \beta^2,$$

for which the solution is

$$Z = \exp\left(-\frac{\alpha}{u_m} \beta^2 z \right),$$

satisfying the boundary condition that $Z = 1$ at $z = 0$ and

$$\frac{\mathrm{d}^2 R}{\mathrm{d}r^2} + \frac{1}{r} \frac{\mathrm{d}R}{\mathrm{d}r} + \beta^2 R = 0,$$

which is a Bessel equation of zero order having a particular integral

$$R = A J_0(\beta r),$$

and to satisfy the boundary condition that $\theta = 0$ at $r = a$ for all values of z greater than 0, we must have

$$J_0(\beta a) = 0,$$

285

i.e. the constant β must be a root of the equation $J_0(\beta a) = 0$. If these values are $\beta_1, \beta_2, \ldots$, etc., the required solution of the Bessel equation is

$$R = A_1 J_0(\beta_1 r) + A_2 J_0(\beta_2 r) + \ldots + A_n J_0(\beta_n r) + \ldots,$$

where A_1, A_2, etc., are constants which are available for fitting the remaining boundary condition that $\theta = f(r)$ or 1 at $z = 0$.

Multiply both sides of the last result by $r J_0(\beta_n r)$ and integrate from 0 to a.

It can be shown that

$$\int_0^a r J_0(\beta_n r) J_0(\beta_m r) \mathrm{d}r = 0 \quad \text{if} \quad n \neq m,$$

and that

$$\int_0^a r J_0^2(\beta_n r) \, \mathrm{d}r = \tfrac{1}{2} a^2 J_1^2(\beta_n a).$$

Hence, if $R = f(r)$ at $z = 0$,

$$\int_0^a r f(r) J_0(\beta_n r) \, \mathrm{d}r = A_n \tfrac{1}{2} a^2 J_1^2(\beta_n a)$$

or

$$A_n = \frac{2}{a^2 J_1^2(\beta_n a)} \int_0^a r f(r) J_0(\beta_n r) \, \mathrm{d}r,$$

and for the special case of uniform temperature at entry, $f(r) = 1$,

$$\int_0^a r J_0(\beta_n r) \, \mathrm{d}r = \frac{a}{\beta_n} J_1(\beta_n a),$$

and therefore

$$A_n = \frac{2}{\beta_n a} \frac{1}{J_1(\beta_n a)},$$

in which case the solution for R is

$$R = 2 \sum_{n=1}^\infty \frac{1}{\beta_n a} \frac{J_0(\beta_n r)}{J_1(\beta_n a)},$$

and therefore the solution for the original partial differential equation is

$$\theta = RZ = 2 \sum_{n=1}^\infty \frac{1}{\beta_n a} \frac{J_0(\beta_n r)}{J_1(\beta_n a)} \exp\left(-\frac{\alpha}{u_m} \beta_n^2 z\right),$$

which is the result quoted in (17.3.5).

For fully developed laminar flow the energy equation from (17.3.4), assuming an undistorted parabolic velocity distribution, is

$$2u_m \left(1 - \frac{r^2}{a^2}\right) \frac{\partial \theta}{\partial z} = \alpha \left[\frac{\partial^2 \theta}{\partial r^2} + \frac{1}{r} \frac{\partial \theta}{\partial r}\right].$$

Following the same procedure as before, it is assumed that θ can be expressed as a product $R(r)Z(z)$. The equation then takes the form

$$\frac{2u_m}{\alpha} \frac{1}{Z} \frac{\mathrm{d}Z}{\mathrm{d}z} = \frac{1}{(1 - r^2/a^2)R} \left[\frac{\mathrm{d}^2 R}{\mathrm{d}r^2} + \frac{1}{r} \frac{\mathrm{d}R}{\mathrm{d}r}\right],$$

and if each side is equated to a constant $-\beta^2$, we will have

$$\frac{1}{Z}\frac{\mathrm{d}Z}{\mathrm{d}z} = -\frac{\alpha}{2u_m}\beta^2,$$

for which the solution is

$$Z = \exp\left(-\frac{\alpha}{2u_m}\beta^2 z\right)$$

and

$$\frac{\mathrm{d}^2 R}{\mathrm{d}r^2} + \frac{1}{r}\frac{\mathrm{d}R}{\mathrm{d}r} + \beta^2\left(1 - \frac{r^2}{a^2}\right)R = 0,$$

for which a series solution may be obtained, taking the place of the Bessel function in the previous case.

EXAMPLES

1. The flow in a pipe-line of diameter D is measured by means of an orifice plate of diameter $0.5D$ with pressure tappings on either side connected to a manometer. The discharge coefficient is 0.62.

Calculate the throat diameter of a venturi tube which would give the same pressure difference at the same rate of flow in the pipe. The discharge coefficient for the venturi tube may be taken as 0.96.

If the efficiency of the diffuser of the venturi tube is 80%, what would be the ratio of the overall pressure drop in the pipe-line caused by the venturi tube to that caused by the orifice plate? The coefficient of contraction for the orifice plate may be taken as 0.64, and it may be assumed that for the measurement of the overall pressure drop the downstream pressure tapping is at a sufficient distance from the orifice plate for uniform flow to be re-established. Neglect pipe friction.

2. A sharp-edged rectangular weir of breadth b, operating with head h measured above the crest, is to be used to measure the flow of a liquid whose coefficient of viscosity is 20 times that of water at $10°$C. and whose specific gravity is 0.96. In order to calibrate the weir experiments are to be made on a scale model using water at $10°$C. as the fluid. Determine the appropriate breadth and working head for the model so that a reliable calibration may be made. What will be the ratio of the two volume flows?

3. CO_2 is removed from a mixture of synthesis gases in a continuous process at 50 atmospheres pressure by scrubbing with water. Power recovery is achieved by letting down the water to atmospheric pressure through a Pelton wheel, which is to drive a direct-coupled generator at 600 r.p.m. Assuming that the maximum jet diameter is limited to one-twelfth of the rotor diameter, determine the maximum flow that could be handled by a single-jet machine under optimum operating conditions. Estimate also the power developed and the hydraulic efficiency of the machine. Assume that the velocity coefficient for the nozzle is 0.95, that the peripheral velocity under optimum conditions is 0.45 of the jet velocity, and that the relative velocity at exit from the wheel is 0.78 of that at entry and inclined at $20°$ to the plane of the wheel.

What would be a suitable material for the rotor of the Pelton wheel?

4. Explain briefly the significance of the 'specific speed' of a centrifugal or axial-flow pump.

A pump is designed to be driven at 600 r.p.m. and to operate at maximum efficiency when delivering 5000 gal. of water per minute against a head of 64 ft. Calculate the specific speed. What type of pump does this value suggest?

A pump, built for these operating conditions, has a measured maximum overall efficiency of 70%. The same pump is now required to deliver water at 100 ft. head. At what speed should the pump be driven if it is to operate at maximum efficiency? What will be the new rate of delivery and the power required?

5. A pipe-line 600 miles long is to be built for conveying crude oil of specific gravity 0·87. The inside diameter of the pipe is to be 12 in. and the quantity of oil to be delivered is 2 million gallons per day. Pipe lengths are available in three different sizes of wall thickness, the heaviest of which is suitable for internal working pressures up to 800 p.s.i. The flow is to be maintained by means of pumping stations spaced at approximately equal intervals along the line.

Neglecting the effect of differences in elevation, and taking a value of 0·0075 for c_f, estimate the total pumping power required and the minimum number of pumping stations. The pressure on the suction side of the pumps is not to be less than 0 p.s.i. (gauge).

6. The arrangements of piping and pump for an acetone storage system is shown diagrammatically in fig. 1. The acetone level in the tank

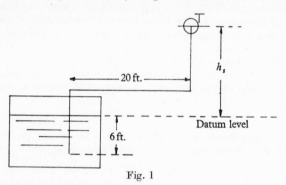

Fig. 1

is maintained constant and the tank is vented to atmosphere. The pump, when transferring 400 gal./min. from the tank, is required to develop a delivery pressure head of 40 ft. measured above the datum level indicated. A single-stage centrifugal pump driven at 1450 r.p.m. is proposed. To avoid cavitation when operating under design conditions with this pump it is necessary to maintain a net positive suction head greater than σH, where $\sigma = 7 \cdot 1 N_s^{\frac{4}{3}} \times 10^{-6}$, $H =$ total head developed by the pump and N_s is the specific speed.

Assuming a diameter of 4 in. for the suction pipe, determine the maximum permissible height h_s of the pump centre-line above the liquid surface level in the tank. For the friction loss in the pipe assume that $c_f = 0.006$ and allow for a loss of one velocity head at each right-angle bend.

Physical properties of liquid acetone at the operating temperature of 20°C.: specific gravity, 0·79; vapour pressure, 185 mm. Hg.

7. Discuss briefly the influence of viscosity on the performance of a centrifugal pump.

A centrifugal pump is to be designed to deliver 3600 gal./min. at 200 ft. head when handling a liquid hydrocarbon product of specific gravity 0·95 and viscosity 2·5 poise. The rotational speed is to be 1450 r.p.m. In order to check the performance of the full-scale pump, a half-scale model is to be tested pumping a light oil of specific gravity 0·90 and viscosity 0·25 poise. At what speed should the model be driven? What should be its delivery head and volume flow at the design point if the specified performance is to be achieved with the full-scale pump?

If the measured power required to drive the model under these conditions is 0·96 h.p., what will be the efficiency of the full-scale pump? Comment on this value.

8. Viscose rayon dope is being pumped at a steady rate along a 4 in. inner diameter pipe 50 ft. long. It is desired to change over from unpigmented to pigmented dope without interrupting the flow. If the flow is laminar, how much pigmented dope will have to be pumped into the pipe before the mean concentration of pigment in the leaving stream rises to 95% of that in the entering stream?

Ignore entrance effects, and assume that pigmented and unpigmented dopes have the same density and viscosity, and behave as Newtonian fluids.

9. In a solvent extraction plant, shown in fig. 2, two immiscible liquids A and B are supplied to a settling tank in which two liquid

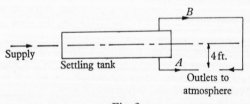

Fig. 2

layers are formed. Tappings in the top and bottom of the tank take off separate streams which are regulated by orifice plates in pipes A and B. Find the diameters of the orifices in lines A and B if the interfacial level in the settling tank is not to vary more than $\pm$ 2 ft. from the centre-line, under the following conditions:

Maximum value of x, the weight fraction of liquid A in the feed: 0·6.

Minimum value of x: 0·4.

Relative densities: liquid A, 1·0;

liquid B, 0·8.

The combined flow rate is fixed at 10,000 lb./hr.

Contraction coefficient for the orifices = 0·6.

Assume that the pipe diameters are much larger than the orifice diameters.

10. Show that the general equation for heat conduction in a solid in which heat is generated internally at a rate $H(x, y, z)$ heat units per second per unit volume is

$$\frac{\partial \theta}{\partial t} = \frac{k}{\rho c}\, \nabla^2 \theta + \frac{H}{\rho c}.$$

Heat is generated uniformly at a constant rate H per unit volume in the material of a long solid cylinder of radius r_1. The cylinder is sheathed with a covering of insulation having an outer radius r_2. The surrounding atmospheric temperature is θ_0. Show that, when steady conditions have been reached, the temperature at the centre of the cylinder is equal to

$$\theta_0 + \frac{Hr_1^2}{2}\Big(\frac{1}{2k_a} + \frac{1}{k_b}\log_e\frac{r_2}{r_1} + \frac{1}{r_2 h}\Big),$$

where k_a = thermal conductivity of the material, k_b = thermal conductivity of the sheath, h = heat-transfer factor between the outer surface of the sheath and the surroundings.

11. A circular pipe of radius r_1 has cooling fins of thickness t and outer radius r_2. The pipe wall is at a temperature θ_1 above the temperature of the surrounding atmosphere. Derive an expression for the temperature distribution in the fins for steady radial conduction of heat. Assume that there is transfer by convection from the sides of the fins with factor h but that there is no flow of heat from the ends of the fins at radius r_2.

Hence show that, for a pipe of radius r_1 with a continuous helical fin of outer radius r_2 and pitch b which may be assumed small compared with r_1, the rate of heat dissipation through the fin per unit length of pipe is given by

$$\frac{2\pi r_1 t}{b}\, \alpha k \theta_1\, \frac{K_1(\alpha r_1)I_1(\alpha r_2) - I_1(\alpha r_1)K_1(\alpha r_2)}{I_0(\alpha r_1)K_1(\alpha r_2) + K_0(\alpha r_1)I_1(\alpha r_2)},$$

where

$$\alpha = \Big(\frac{2h}{kt}\Big)^{\frac{1}{2}}.$$

It may be assumed that the modified Bessel functions obey the relations

$$\frac{\mathrm{d}I_0(x)}{\mathrm{d}x} = I_1(x), \quad \frac{\mathrm{d}K_0(x)}{\mathrm{d}x} = -K_1(x).$$

12. A gas to liquid heat exchanger is constructed with alternate gas and water channels and with internal fins in the gas passages arranged as shown in the cross-sectional view of fig. 3. Dimension b is large compared with a. Assuming a uniform temperature T_0 for the walls of the

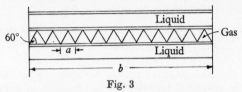

Fig. 3

channels and an average heat-transfer factor h on the gas side, show that the total heat flux per unit length for one gas channel is given closely by

$$Q = 2bh(T_1 - T_0)\left[1 + \frac{4}{\lambda a}\tanh\frac{\lambda a}{2}\right],$$

where $T_1 =$ mean temperature of the gas, $\lambda = \sqrt{(2h/kt)}$, $t =$ fin thickness, $k =$ thermal conductivity of the fin metal.

13. Heat is generated at a uniform rate H per unit volume in a flat plate with surface dimensions which are large compared with its thickness l. Heating is commenced at time $t = 0$ with the plate at a uniform temperature θ_0. The surface temperatures are maintained at θ_0.

Show that the ultimate steady temperature is given by

$$\theta_f - \theta_0 = \frac{H}{2k}x(l - x),$$

where the surfaces are given by the planes $x = 0$ and $x = l$. Hence by assuming that

$$\theta = \theta_f + \phi(x, t),$$

show that at any time t

$$\theta - \theta_0 = \frac{H}{2k}\left[x(l - x) - \frac{8l^2}{\pi^3}\sum_{n=1,3}^{\infty}\frac{1}{n^3}\exp\left(-\frac{k}{\rho c l^2}n^2\pi^2 t\right)\sin n\pi\frac{x}{l}\right].$$

14. Establish the equation

$$\frac{\partial\theta}{\partial t} = \alpha\operatorname{div}\operatorname{grad}\theta, \quad \text{where} \quad \alpha = \frac{k}{\rho c}.$$

A stirred tank contains liquid z at its freezing temperature θ_1. Solid z is formed on the top of the liquid by reducing the upper surface temperature to zero at time $t = 0$ and subsequently maintaining it there. Throughout the solidification process convection currents maintain the liquid at θ_1. At time t the thickness of the solid layer is b. The latent heat of melting of z is l.

Show that the boundary conditions and the equation of heat conduction in the solid are satisfied if

$$b = 2\beta \sqrt{(\alpha t)},$$

$$\theta = A \operatorname{erf} \frac{x}{2\sqrt{(\alpha t)}},$$

where β, A are constants, and x is the distance from the upper surface of the solid. Hence show that

$$\beta e^{\beta^2} \operatorname{erf} \beta = \frac{c\theta_1}{\sqrt{\pi}\, l}.$$

15. In a counter-flow waste-heat recuperator, a stream of hot waste gas gives up heat to a stream of cold air which is flowing in the opposite direction. The mass flow rate of the waste gas is w_g and its mean specific heat c_g, while the corresponding quantities for the air are w_a and c_a respectively. The effective surface area for heat transfer from the hot gas to the cold air is S, and the mean overall heat-transfer factor is h.

Show that the efficiency of the recuperator, defined as the ratio of the heat actually gained by the air to the heat that would be gained if the air attained the inlet temperature of the hot waste gas, is given by

$$1 - \left(1 - \frac{w_g c_g}{w_a c_a}\right) \Big/ \left(1 - \frac{w_g c_g}{w_a c_a}\, e^{-\alpha}\right),$$

where

$$\alpha = hS\left(\frac{1}{w_g c_g} - \frac{1}{w_a c_a}\right).$$

A recuperator is designed to achieve an efficiency of 60% when heating 60 lb./sec. of air ($c_p = 0.240$) with a stream of 50 lb./sec. of hot waste gas ($c_p = 0.250$). By what factor would the heating surface have to be increased to raise the efficiency to 70%? Assume that the value of h remains unchanged.

16. Nitrogen is compressed from 1 atmosphere and 15° C. to a pressure of 6 atmospheres in a multi-stage turbo-compressor having an isentropic efficiency of 75%. It is required to cool the gas to 30° C. before passing it on to the next compressor.

A single-pass shell and tube heat exchanger is to be employed for this purpose using water outside the tubes at 10° C. as the cooling medium. The mass velocity of the gas through the tubes is to be 15 lb./sq.ft.sec. and the inside diameter of the tubes is $\frac{1}{2}$ in.

Considering only the resistance to heat transfer on the gas side, calculate the tube length required. Assume that $Nu = 0.023 Re^{0.8} Pr^{0.4}$.

17. Two tests on a shell and tube heat exchanger with oil on the tube side, and condensing steam at atmospheric pressure on the shell side, gave the following results:

Oil flow (lb./hr.)	197	19,650
Steam condensed (lb./hr.)	15·5	530

For the oil: $\rho = 47, \mu = 2\cdot0, k = 0\cdot06, c_p = 0\cdot50$, in lb. ft.hr. °F. units.

The exchanger was 7 ft. long, contained 30 tubes of $\frac{1}{2}$ in. inner diameter, and the shell was 6 in. outer diameter. There were convection losses to the atmosphere from the shell, for which $h = 2$ B.TH.U./sq. ft.hr. °F. The inlet and atmospheric temperatures were both 60° F.

Estimate, for each test, the value of Nu and the pressure drop across the exchanger.

18. A CO_2 compressor handles a flow of 2 lb./sec. with a pressure and temperature at suction of 14·7 lb./sq.in. absolute and 50° F. The delivery pressure is 80 lb./sq.in. and the compression may be assumed isentropic. The gas from the compressor is to be cooled to 86° F. using a longitudinal flow heat exchanger with two passes on the gas side, CO_2 inside the tubes, and water at a mean temperature of 68° F. outside the tubes. It is proposed to use steel tubes of $\frac{3}{4}$ in. outside diameter × 18 s.w.g. wall thickness (0·654 in. inner diameter).

If the permissible overall pressure drop on the gas side is limited to 1 lb./sq.in., determine the number of tubes required and the distance between tube plates.

For heat transfer on the gas side assume that $St = 0\cdot043Re^{-\frac{1}{4}}$ and on the water side that $h = 500$ B.TH.U./hr.sq.ft. °F. For the friction loss assume that $c_f = 0\cdot080Re^{-\frac{1}{4}}$, but allow also for pressure drop at inlet to and outlet from the tubes by adding a total of $\frac{3}{2}\rho u_m^2$ to the pressure drop caused by skin friction alone, where u_m = mean velocity in the tubes.

19. A heat exchanger is to be designed to heat a flow of 500 gal./min. of heavy oil from 25° to 75° C. using flue gas which is available at an entry temperature of 400° C. and total mass flow rate 40 lb./sec. The exchanger is to be in the form of a bank of tubes, of 1 in. inner diameter and $1\frac{1}{4}$ in. outer diameter, with 10 ft. between tube plates. Tube rows are to be staggered and there is to be a clear space of 1 in. between adjacent tubes in a row. Flue gas is to flow across the tube bank and the pressure drop is limited to 2 in. water. The velocity of the oil inside the tubes is to be approximately 2 ft./sec.

Determine a suitable tube arrangement for the heat exchanger. Estimate also the pumping power required to maintain the oil flow, allowing an additional 10 ft. head for other losses in the pipe-lines. Specify a suitable type of pump for this application.

For heat transfer on the gas side take $Nu = 0{\cdot}35Re^{0{\cdot}6}Pr^{0{\cdot}7}$ and on the oil side take $Nu = 1{\cdot}62\left(RePr\dfrac{d}{L}\right)^{\frac{1}{3}}$. For the pressure drop on the gas side assume $\Delta p = 1{\cdot}5n\rho v^2 \times 10^{-3}$ in. water, where $n =$ number of tube rows and $v =$ mean velocity in the space between adjacent tubes.

Properties of the oil: specific gravity, $0{\cdot}96$; specific heat, $0{\cdot}48$; viscosity, $0{\cdot}5$ poise at 50° C.; thermal conductivity, $0{\cdot}080$ c.h.u./ft.hr. $^\circ$C.

Assume properties of air for the flue gas.

20. A fluid A initially at temperature T_1 is cooled in a heat exchanger to T_2 by the counter-current flow of a fluid B initially at temperature t_1 and finally at t_2.

If A has a specific heat C and is flowing at a total mass flow rate W, the corresponding values for B are c and w, and the overall transfer coefficient U based on an effective area α is assumed constant, show that

$$\frac{1-R}{R}(T_1 - T_2) = \Delta\theta_1 - \Delta\theta_2 = \Delta\theta_1\left\{1 - \exp\left(-(1-R)\frac{U\alpha}{wc}\right)\right\},$$

where $\qquad R = \dfrac{wc}{WC}, \quad \Delta\theta_1 = T_2 - t_1 \quad$ and $\quad \Delta\theta_2 = T_1 - t_2.$

If the exchanger is changed to a series-parallel arrangement, the cooling fluid being divided into two equal streams each flowing counter-currently through half the exchanger as indicated diagrammatically in fig. 4, show that the temperature T_c of A at its exit from the first section is given by

$$T_c - t_1 = \sqrt{\{(T_1 - t_1)(T_2 - t_1)\}}.$$

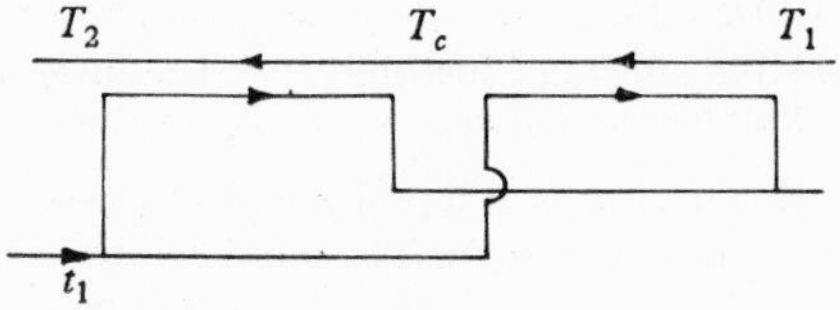

Fig. 4

A double-pipe heat exchanger is to be used to cool 5000 lb./hr. of a lubricating oil distillate from 450 to 350°F. by using 20,000 lb./hr. of crude oil initially at 300°F. If a series-parallel arrangement is to be used, the crude oil being the divided stream and flowing in the inner pipe, estimate the effective total tube length of the exchanger and the corresponding pressure drops in each section.

Assume

$$\frac{h}{cG}(Pr)^{\frac{2}{3}} = \frac{c_f}{2} = 0{\cdot}023Re^{-0{\cdot}2},$$

and neglect all thermal resistances other than those of the liquid streams.

The following mean values of the relevant physical properties all expressed in lb., ft., hr., B.TH.U., °F. units may be used:

Lubricating oil distillate: $\mu = 7 \cdot 0$, $k = 0 \cdot 07$, $c_p = 0 \cdot 62$, $\rho = 48$.

Crude oil: $\mu = 2 \cdot 0$, $k = 0 \cdot 075$, $c_p = 0 \cdot 60$, $\rho = 47$.

Dimensions of exchanger tubes:

Inner tube: $d_i = 1 \cdot 38$ in., $d_o = 1 \cdot 66$ in.

Outer tube: $d_i = 2 \cdot 07$ in.

21. A long push-rod in an oil-operated control system is located in a tube with small radial clearance δ. A pressure difference Δp is maintained between the ends of the tube, and there is a steady leakage of oil in the annular space between the push-rod and the tube. Show that, if the push-rod is displaced to one side so that it touches the wall of the tube, the rate of leakage of the oil will be approximately $\frac{5}{2}$ times the rate of leakage which would occur with the push-rod in a central position. Assume laminar viscous flow of the oil.

22. Two immiscible fluids A and B of appreciably differing density are flowing co-currently through a closed channel of rectangular cross-section of depth $2a$ and breadth b. It is observed that the flow of both fluids is laminar and that over the test length each forms a uniform layer of depth a. If the axis of the channel is horizontal and b is very much greater than a show that the volume flow of A is given by

$$\frac{\Delta P a^3 b}{12 \mu_A l} \frac{7 \mu_A + \mu_B}{\mu_A + \mu_B},$$

where ΔP is the pressure drop over l and μ_A and μ_B are the viscosities of A and B respectively.

Neglect the possible effects of instability at the interface.

23. A cross-sectional view of a liquid metal electromagnetic pump is shown diagrammatically in fig. 5. The channel in which the liquid

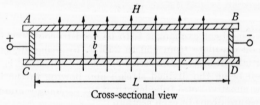

Cross-sectional view

Fig. 5

metal flows is of narrow rectangular cross-section $ABCD$. The sides AB and CD of this channel are made of insulating material, while the sides AC and BD are conductors. The dimension L is large compared with b. A potential difference V is maintained between the two conducting walls causing a transverse electric current of density i to flow. A magnetic field H is also maintained across the channel in a direction perpendicular both to the velocity of flow of the metal and to the electric current. The velocity at a distance y from the centre line (i.e. $\frac{1}{2}b + y$ from a wall) is u. All quantities involved are measured in c.g.s. or the corresponding electromagnetic units.

By considering the motion with the stream of a current filament show that

$$i = \frac{V}{\sigma L} - \frac{uH}{\sigma},$$

where σ = specific resistance of the liquid metal.

Hence show that the velocity distribution is given by

$$u = \left(\frac{V}{LH} - \frac{\sigma}{H^2} \frac{dp}{dx} \right) \left[1 - \frac{\cosh \lambda y}{\cosh \lambda(\frac{1}{2}b)} \right],$$

where $\lambda = H/\sqrt{(\sigma\mu)}$, μ = viscosity of the liquid metal and dp/dx = pressure gradient in direction of flow.

Treat the problem as one of two-dimensional laminar viscous flow, i.e. allow for zero slip at the walls AB and CD but ignore the effects of the walls AC and BD.

24. Fluid flows from a tank to a long straight pipe of radius a via a short well-rounded entrance piece. Along the entry length, i.e. that part of the pipe in which the actual velocity distribution differs sensibly from the final steady distribution, the flow is laminar and may be assumed to approximate to that indicated in fig. 6.

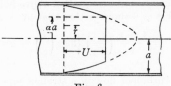

Fig. 6

Show that the rate of transfer of momentum across any cross-section of the pipe is $\pi a^2 [\frac{1}{3}(1 + 2\alpha^2)\rho U^2]$, and that the entry length x, i.e. the distance from $\alpha = 1$ to $\alpha = 0$, is given by

$$\frac{x}{a} = z \frac{\rho \bar{U} a}{\mu},$$

where $\bar{U}$ is the mean velocity $= \frac{1}{2}(1 + \alpha^2)U$ and

$$z = \int_0^1 \frac{(1 - \beta)(3 - 2\beta)}{6(1 + \beta)^2} \, d\beta = 0 \cdot 127.$$

25. A porous flat plate forms one boundary of a non-turbulent air stream. The undisturbed air is at constant pressure and flows with a uniform velocity U_0 parallel to the surface of the plate. Air is sucked into the plate so as to induce a uniform normal velocity U_s at the surface where U_s is small compared with U_0. Fig. 7 shows diagrammatically the boundary layer formed close to the surface.

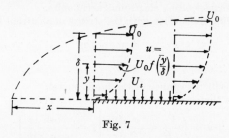

Fig. 7

Assuming that the boundary layer has a finite thickness δ and that the longitudinal component of the air velocity is given by an equation of the form $u = U_0 f(\eta)$, where $\eta = y/\delta$, show by considerations of continuity and momentum that

$$\frac{\tau}{\rho U_0^2} = \frac{U_s}{U_0} + \frac{\mathrm{d}\delta}{\mathrm{d}x} \int_0^1 f(1-f)\mathrm{d}\eta,$$

where τ is the local surface shear stress.

If the flow in the boundary layer is non-turbulent and it is assumed that $f = \sin \pi/2 \, \eta$, show that δ approaches a limiting value given by

$$\delta_L = \frac{\pi}{2} \frac{\mu}{\rho U_s}.$$

Assuming that the momentum equation above holds from the leading edge at which point $\delta = 0$ show further that $(\delta_L - \delta)/\delta_L = \alpha$ when

$$\frac{\rho U_0 x}{\mu} = \frac{4 - \pi}{4} \left(\frac{U_0}{U_s}\right)^2 \left\{ \log_e \left(\frac{1}{\alpha}\right) + \alpha - 1 \right\}.$$

26. A fluid is in turbulent flow in a straight pipe of radius a. In the turbulent region the mean axial velocity u at radius r is given by

$$u = u_0 \left[1 - \left(\frac{r}{a}\right)^2 \right]^n.$$

If the velocity distribution close to the wall is dependent on τ, ρ, μ and the distance from the wall y only and not on a, show that the skin-friction coefficient c_f is given by an equation of the form

$$c_f = \alpha Re^{-\frac{2n}{1+n}}.$$

It may be assumed that the thickness δ of the laminar sub-layer is small and that its effect can be neglected in evaluating the bulk flow.

If experimentally $c_f = 0 \cdot 079 Re^{-\frac{1}{4}}$, make a first estimate of the magnitude of δ/a.

27. Diffusion is occurring in a two-component gaseous system. The local mean molecular weight is M and the local velocity of the mixture, defined as the local mass velocity divided by the local density, is $\mathbf{v}$.

Obtain expressions for the mass outflow and for the molal outflow from an element of volume, and show that

$$\frac{\partial \log_e M}{\partial t} + \mathbf{v} \, . \, \mathrm{grad} \, \log_e M = - \, \mathrm{div} \, \mathbf{v} = \mathscr{D} \, \mathrm{div} \, \mathrm{grad} \, \log_e M.$$

It may be assumed that $c_2 N_1 - c_1 N_2 = - c \mathscr{D}(\partial c_1 / \partial z)$ and that the pressure is sensibly constant.

28. Simultaneous heat and mass transfers are occurring across a plane laminar gas film of thickness δ. The gas consists of two components C and D with molal specific heats C_p' and C_p'' respectively, and the components form an ideal gas mixture. At any one point the mole fraction of C is y, the density is ρ, the temperature is θ and the mean molecular weight is M. The two boundaries of the film are denoted by a and b, and the transfers are measured as positive from b to a. Show that per unit area per unit time the total molal transfer is

$$N = \frac{\beta}{\delta} \log_e \frac{y_a - z}{y_b - z},$$

and that the conduction heat transfer at the boundary a is

$$q = \frac{k}{\delta} (\theta_b - \theta_a) \frac{\alpha}{1 - \mathrm{e}^{-\alpha}},$$

where $\beta = \rho \mathscr{D}/M$ and k are assumed sensibly constant through the film, $z = $ the mole fraction of the transfer N which is due to component C; and $\alpha = N C_p (\delta/k)$, where

$$C_p = z C_p' + (1 - z) C_p''.$$

It may be assumed that $c_2 N_1 - c_1 N_2 = - c \mathscr{D}(\partial c_1 / \partial x)$.

A turbulent two-component gas stream is flowing with mass velocity G through a pipe. Mass and heat transfers are occurring to the walls. Assuming that the transfers can be evaluated by using effective film thicknesses connected by the relations

$$\frac{d}{\delta} = \frac{d}{\delta_m} Sc^{-\frac{1}{3}} = \frac{d}{\delta_h} Pr^{-\frac{1}{3}} = Re \, \frac{c_f}{2},$$

show that

$$\frac{\beta}{\delta_m} = \frac{G}{M} Sc^{-\frac{2}{3}} \frac{c_f}{2},$$

and that

$$\frac{k}{\delta_h} = \frac{G C_{pf}}{M} Pr^{-\frac{2}{3}} \frac{c_f}{2},$$

where Sc, Pr and C_{pf} are evaluated at the mean film conditions.

29. A rocket is propelled by the combustion of a mixture of ethyl alcohol and liquid oxygen, the products of combustion passing through a convergent-divergent thrust nozzle. The rate of supply of the two fuels is such that a pressure of 20 atmospheres is maintained in the combustion chamber, and the temperature at this point is approximately 3000° K. The throat diameter of the thrust nozzle is 1·32 ft. Assuming that the products of combustion may be treated as a perfect gas with the mean molecular weight 20 and $\gamma = 1·25$, calculate:

(a) the correct diameter of the exit so that full thrust may be developed at ground level;
(b) the mass flow of fuel;
(c) the thrust.

30. Show that, for the isentropic flow of a perfect gas from a reservoir at pressure p_0, the relation between the Mach number M and the pressure p at any point is given by

$$\frac{p_0}{p} = \left[1 + \frac{\gamma - 1}{2} M^2\right]^{\gamma/(\gamma - 1)}.$$

Show also that for isentropic flow in a correctly designed convergent-divergent nozzle, operating with sonic velocity at the throat,

$$\frac{A_t}{A} = \left(\frac{\gamma + 1}{2}\right)^{1/(\gamma - 1)} \left[\frac{\gamma + 1}{\gamma - 1} \left\{\left(\frac{p}{p_0}\right)^{2/\gamma} - \left(\frac{p}{p_0}\right)^{(\gamma + 1)/\gamma}\right\}\right]^{\frac{1}{2}},$$

where p is the pressure at a point where the cross-sectional area is A, and A_t is the throat area.

Outline the derivation of the following relations connecting the pressures and Mach numbers on either side of a normal shock wave

$$\frac{p_2}{p_1} = \frac{2\gamma M_1^2 - (\gamma - 1)}{\gamma + 1} \quad \text{and} \quad M_2^2 = \frac{(\gamma - 1)M_1^2 + 2}{2\gamma M_1^2 - (\gamma - 1)},$$

where subscript 1 refers to conditions immediately upstream and 2 to conditions immediately downstream from the shock.

A gas having a value of $\gamma = 1·3$ flows from a reservoir at pressure p_0 through a convergent-divergent nozzle having throat area A_t and exit section area $2A_t$. Under certain flow conditions a normal shock wave is situated at a section where the flow area is $1·2A_t$. Calculate the back pressure in terms of p_0. What would be the correct back pressure for which the flow in the divergent portion would just remain supersonic with absence of shock waves?

31. Derive the following expression for the compressible flow of a gas through a uniform pipe with skin-friction coefficient c_f:

$$\int_1^2 \rho \mathrm{d}p + G^2 \log_e \frac{\rho_1}{\rho_2} + 2c_f \frac{L}{D} G^2 = 0,$$

where G = mass velocity, L = length of pipe between sections 1 and 2, D = pipe diameter, ρ = density of the gas.

Ethylene is to be pumped along a 6 in. inner diameter pipe for a distance of 5 miles at a total mass flow rate of 2 lb./sec. The delivery pressure at the end of the pipe is to be 2 atmospheres and the flow may be assumed isothermal at temperature 60°F. At the inlet end the gas is to be compressed from 1 atmosphere and 60°F. up to the necessary pressure and then passed through a cooler before entry to the pipe-line. Assuming isentropic compression, estimate the pumping power required. For this calculation the superheated ethylene may be treated as a perfect gas with mean $c_p = 0 \cdot 415$. For skin friction in the pipe take $c_f = 0 \cdot 003$.

32. A rapid-flowing stream in an open channel changes over to tranquil flow at a certain point through a hydraulic jump. The channel is of wide rectangular cross-section and the volumetric rate of flow is Q per unit width of stream. Derive an expression connecting the depths of flow on either side of the jump.

Water is flowing under a sluice gate from a reservoir in which the depth is 10 ft., and discharges into a long open channel of uniform rectangular cross-section and bottom slope s_0. The height of the opening under the sluice gate is 3 ft., and for free outflow a coefficient of contraction of 0·62 and a velocity coefficient of 0·96 may be assumed. For uniform flow in the open channel, a value of $c_f = 0 \cdot 0033$ may be taken. Estimate the minimum value of the bottom slope of the channel for which the flow below the sluice gate will remain free.

Describe briefly how you would predict the location of the hydraulic jump, assuming free outflow from the sluice gate into a very long channel of mild slope.

33. A liquid is in steady uniform flow along an open channel having a small angle of bottom slope α. The depth of the stream is λ and the motion is fully turbulent. There is a suspension of small solid particles of approximately uniform size in the stream. On the basis of the assumptions listed below derive the following expression for the distribution of solid matter in the stream:

$$\frac{n}{n_0} = \left[\frac{y}{y_0} \frac{(1 - y_0/\lambda)}{(1 - y/\lambda)} \right]^{- \frac{v_g}{k(\lambda g \sin \alpha)^{\frac{1}{2}}}},$$

where y = distance measured perpendicular to channel floor, n = particle density at level y, n_0 = particle density at specified level y_0, v_g = terminal settling velocity of the particles under gravity, k = constant in the mixing length formula below.

Assume that

(i) the particles, while maintaining their settling velocity relative to the fluid are subject to the same turbulent movements as the fluid,

(ii) the particles do not affect the momentum transfer appreciably,

(iii) the Prandtl mixing length can be represented by $l = ky(1 - y/\lambda)^{\frac{1}{2}}$.

34. Derive the following expression for the pressure drop occurring in the viscous flow of a fluid through a bed of spherical particles of mixed sizes:

$$\Delta p = k \frac{(1 - \varepsilon)^2}{\varepsilon^3} \frac{\mu u_m L}{D_{vsm}^2},$$

where $k =$ a constant, $\varepsilon =$ fraction void, $u_m =$ mean approach velocity, $L =$ bed height, $D_{vsm} =$ volume-surface mean diameter $= \dfrac{\Sigma D_p^3}{\Sigma D_p^2}$, $D_p =$ diameter of an individual particle.

Assuming a log-probability distribution of particle diameter, with a standard deviation σ, show that the pressure drop can be represented by

$$\Delta p = k \frac{(1 - \varepsilon)^2}{\varepsilon^3} \frac{\mu u_m L}{D_m^2} e^{\sigma^2},$$

where $D_m =$ median diameter for mass distribution as determined by sieving measurements.

35. A steam-driven double-acting reciprocating pump is employed to feed a suspension of a homogeneous compressible sludge to a plate and frame filter press. The steam supply is such that the maximum discharge pressure attainable by the pump is 50 lb./sq.in. gauge with the suction at atmospheric pressure.

During a test, the pump was maintained at a constant maximum pumping rate until the discharge pressure of 50 lb./sq.in. was reached. The filtering process then continued at this constant pressure. The amount of filtrate was measured, starting from the time when the first filtrate left the press. The results obtained are given in the following table:

Time (min.)	10	15	18	20	25	30	40	50	60
Weight of filtrate (lb.)	500	750	900	1000	1225	1414	1732	2000	2235
Pressure on pump (lb./sq.in.)	$1\frac{1}{2}$	12	29	50	50	50	50	50	50

If the cleaning of the filter necessitates a complete shut-down for 30 min., what is the maximum daily (24 hr.) production of filtrate?

In an effort to improve the daily capacity, it is proposed to install a second pump identical with the first. Estimate the percentage increase in production (*a*) if the two pumps are operated in parallel, and (*b*) if the first pump discharges into the suction pipe of the second.

36. A filter press of 200 sq.ft. filtering surface is employed to filter a homogeneous suspension of a compressible sludge at maximum capacity with the pump employed, and a test run under typical operating

conditions (the filter discharging to a trough at atmospheric pressure) gave the following results:

Time (min.)	10	15	20	24	25	40	60
Volume of filtrate (gal.)	600	900	1200	1440	1498	2204	2880
Pressure in pump (lb./sq.in. gauge)	10	23	43	60	60	60	60

The cleaning of this filter necessitates a complete shut-down for 30 min.

A continuously operating rotary vacuum filter is also available. This has a diameter of 6 ft. and a width of 6 ft. and the internal segments are arranged so that one-sixth of the total filtering surface is effective at any time. The filter is fitted with a variable speed control, and a vacuum pump maintains an absolute pressure of $6 \cdot 7$ lb./sq.in. inside the drum, and the slurry trough is at atmospheric pressure.

At what speed must the vacuum filter be rotated in order to equal the maximum daily (24 hr.) production of the plate and frame filter? Neglect all resistances except that of the filter cake.

37. A gas cooler consists of a cylindrical vessel filled with a granular material. The vessel is of radius a, axial length L and has its cylindrical wall maintained at a low temperature θ_0. The gas entry temperature is θ_1 and its volume rate of flow is $\pi a^2 u_m$. Longitudinal heat transfer in the cooler may be assumed to occur by convection only and radial transfer by conduction only, the effective radial conductivity being k_e.

Assuming that the effective longitudinal velocity u_m is sensibly constant throughout the cooler, derive a differential equation for determining the temperature θ at any point in the cooler and hence show that θ_m, the mean gas temperature at exit, is given by

$$\frac{\theta_m - \theta_0}{\theta_1 - \theta_0} = 4 \sum_1^\infty \frac{1}{\alpha_i^2} \exp\left(-\beta \frac{\alpha_i^2}{a^2} L\right),$$

where $\beta = k_e / \rho c_p u_m$ and a_i is the ith root of $J_0(\alpha) = 0$.

Note:

$$\frac{2}{a^2 J_1^2(\alpha_i)} \int_0^a r J_0\left(\frac{\alpha_i r}{a}\right) J_0\left(\frac{\alpha_k r}{a}\right) dr = \delta_{ik},$$

$$\int_0^a r J_0\left(\frac{\alpha_i r}{a}\right) dr = \frac{a^2}{\alpha_i} J_1(\alpha_i).$$

38. Derive the Nusselt expression for heat transfer in film condensation of a vapour on a smooth vertical surface

$$Nu = \frac{hx}{k} = 0 \cdot 943 \left(\frac{l \rho^2 g x^3}{k \Delta \theta \mu}\right)^{\frac{1}{4}},$$

where l = latent heat, x = height of surface, $\Delta\theta$ = mean temperature difference, k = thermal conductivity of the liquid, μ = viscosity of the liquid, ρ = density of the liquid.

Discuss the reasons why there is sometimes an appreciable difference between the figures given by this expression and the measured values.

For condensation on a vertical tier of n horizontal tubes, the following expression may be used:

$$Nu = \frac{hd}{k} = 0 \cdot 72 \left(\frac{l \rho^2 g d^3}{nk \Delta \theta \mu} \right)^{\frac{1}{4}}.$$

Calculate the weight of condensate per hour for one vertical tier of 12 tubes of 1 in. outside diameter and 10 ft. length, if ammonia vapour at 138 p.s.i. and 74° F. is condensing on the outside of the tubes. The tubes are water-cooled and the wall temperature is 62° F.

Properties of liquid ammonia at 68° F.: $k = 0 \cdot 286$ B.TH.U./hr.ft. °F.

$$\rho = 38 \cdot 1 \text{ lb./cu.ft.}$$

$$\mu = 0 \cdot 52 \text{ lb./ft.hr.}$$

Latent heat of ammonia at 74° F. 505 B.TH.U./lb.

39. A pure saturated vapour is being condensed on a vertical plate. At a given section the condensate film is turbulent. If it is assumed (a) that the velocity profile consists of a laminar sub-layer and a turbulent region in which $u^* = u \bigg/ \sqrt{\dfrac{\tau_0}{\rho}}$ is effectively constant, (b) that the laminar sub-layer provides the major resistance to heat transfer, and (c) that the laminar sub-layer thickness is small compared with the total film thickness, show that

$$\frac{h}{k} \left(\frac{\mu^2}{\rho^2 g} \right)^{\frac{1}{3}} = A \left(\frac{4 \Gamma}{\mu} \right)^{\frac{1}{3}},$$

where h is local heat-transfer coefficient; k, μ and ρ relate to the condensate properties; Γ is the total mass flow of condensate per unit width of plate; and A is a constant.

If the plate surface temperature is constant and the film is turbulent over most of its length, show that a similar equation holds for the average heat-transfer coefficient.

40. A three-effect evaporator is designed to handle 150 tons of salt per day. Saturated brine at 60° F. is fed in parallel to each effect. Dry saturated steam at 30 p.s.i. absolute pressure is supplied to the first effect and a vacuum of 27 in. Hg is maintained by the condenser. The evaporator bodies and heating elements are identical in the three effects. Salt slurry is removed and pumped to a rotary filter, the filtrate being re-cycled via a transfer tank to the brine feed.

Assuming overall heat-transfer factors of 250, 200 and 150 B.TH.U./hr.sq.ft. °F. and boiling-point elevations of 17, 15 and 13° F. in the first, second and third effects respectively, estimate the required heating surface for one evaporator body and the total steam demand. Determine, also, the pressure in the vapour space of the second effect.

If there are two such evaporator installations in a factory, if process steam demand for other purposes amounts to 5000 lb./hr., and if 2500 kW. of power are required, suggest a suitable arrangement for combining the power and process steam demand with a single boiler installation. Specify the boiler steam conditions and evaporative rate.

Saturated brine at 60° F. contains 35·9 lb. of salt per 100 lb. of water and its mean specific heat over the temperature range involved may be taken as 0·786.

INDEX

absolute pressure, 2
acceleration of a fluid, 16, 149
angular momentum, 37, 44, 46
approach velocity, 253
average deviation, 243

Batchelor, 187
Bernoulli equation, 17, 23, 32, 275
Blair, 80
Blasius, 68, 162
boiling liquids, 267
boundary layers, 3, 160 f., 202, 209, 283
boundary-layer thickness, 162, 203
broad-crested weir, 29, 235
Buckingham, 56

Carslaw, 117
cascade of vanes, 47
cavitation, 92
centrifugal compressor, 99
centrifugal pump, 83
Chezy, 78
choked nozzle, 52, 221
circulation, 49, 277
coefficient of contraction, 26
coefficient of viscosity, 6, 150
Colburn, 271
compressible flow, 49, 52, 219 f.
compressors, 95 f.
condensation, 265
conduction, 11, 104 f.
continuity, 14 f., 191
contraction coefficient, 26
control surface, 14, 30, 35
convection, 12, 17, 129, 139, 208
convective differential, 17
correlation coefficient, 176
counter-flow heat exchanger, 121
critical depth, 234
critical pressure ratio, 51, 221
cyclone collector, 250

density, 2
diffuser, 25, 99
diffusion, 12, 188 f.
diffusivity, 12
dimensional analysis, 55 f., 88, 102, 128 et seq.

discharge coefficient, 25, 26
displacement thickness, 162
draft tube, 47
drag, 56, 62, 157
dynamical similarity, 59, 151, 193, 200

eddy viscosity, 178
ejector, 40
energy equation, 30 f., 85, 198 f.
energy gradient, 74
enlargement of a pipe, 38
enthalpy, 11, 31, 199
error function, 243
Euler equation, 275
evaporators, 170

fan, 61
Fanning equation, 65
Fanno line, 225, 229
filtration, 257
finned surfaces, 113, 115
first law of thermodynamics, 8, 32, 199
Fishenden, 13, 127, 135, 140
flow measurement, 22 f.
fluidized solids, 260
force coefficient, 57, 59
forced convection, 129, 208 f.
forced vortex, 20
form drag, 158
fouling factor, 119
Fourier conduction equation, 105
Francis turbine, 45
free convection, 139
free surface, 62
free vortex, 22
friction, 10, 31, 34, 199
friction coefficient, 65
friction velocity, 72
Froude number, 63, 153

gases, 1
gauge pressure, 2
Gaussian distribution, 243
Gibson, 54
glands, 82
Goldstein, 80, 159, 173, 187
Grashof number, 140

307